The Pulse of the Earth

BY

J. H. F. UMBGROVE
PROFESSOR OF GEOLOGY AT DELFT
HOLLAND

With 6 partly coloured plates, 94 textfigures and 2 tables

SPRINGER-SCIENCE+BUSINESS MEDIA, B.V. 1942

Additional material to this book can be downloaded from http://extras.springer.com

ISBN 978-94-017-6445-2 ISBN 978-94-017-6568-8 (eBook)

DOI 10.1007/978-94-017-6568-8

Originally published by Martinus Nijhoff, The Hague, Holland in 1942

Softcover reprint of the hardcover 1st edition 1942

To
A. C. Umbgrove-Gordon
with sincere feelings of gratitude

CONTENTS

LIST OF ILLUSTRATIONS [1]

TEXTFIGURES

---

[1] If no source is mentioned the illustrations are new or from recent papers by the author.

Figures

Figures

Plates and tables at the end of the book

PREFACE

Problems of current interest relating to the earth's physical history will be discussed in this volume. With the exception of Chapters I and VIII, the topics are adapted from lectures delivered in the years 1939–1941, most of them under the auspices of the Geophysical Section of the Geological and Mining Society for the Netherlands and its Colonies („Geologisch-Mijnbouwkundig Genootschap voor Nederland en Koloniën"), *also before the Geological Society of Leyden* („Leidsche Geologische Vereeniging") *and the Diligentia Society at The Hague* („Maatschappij Diligentia"). *I am much indebted to the first named Society who kindly offered a grant for the publication of this book.*

Each chapter constitutes a subject in itself, but the sequence I have chosen will, I hope, show and explain the deeper correlation of several terrestrial processes which, at first sight, appear to be heterogenous.

The geologist follows the altering face of the earth, the oscillations of the sea-level, the pulsation of folding and mountain-building, the periodicity of the ice-ages, the rhythmical cadence of Life. Just as the physician will draw his conclusions from outward symptoms when examining his patient, the geologist will try to discover the deeper significance of the sequence of observed phenomena by feeling the pulse of the earth.

A few fundamental geological terms have been explained in Chapter I. Geologists and older students will, I trust, forgive me for handling facts that no longer represent terra incognita *to them, realizing that younger undergraduates, biologists and other readers will be glad to dispose of a concise synopsis concerning matters which, in the following chapters, are presupposed to be generally known. On the other hand the Appendix, consisting of a dry and monotonous compilation and discussion of data, is only intended for geologists who desire to examine the fundamental points laid down in plates 1–6 and Chapters II and III.*

It would necessitate too long a list if I were to mention all the names of those who, in one way or another, have generously given assistance or contributed information, to all of whom I hereby wish to express my gratitude. I owe special thanks to my colleagues Dr. J. H. Oort, Dr. H. J. Lam and Dr. B. G. Escher of Leyden, Dr. Ph. H. Kuenen of Groningen, Dr. J. A. A. Mekel and Dr. F. A. Vening Meinesz of Delft for their illuminating discussion or critical reading of parts of the manuscript. I found Mr. J. L. van Houten, B.A., prepared to undertake the heavy task of translating the Dutch manuscript into English. To him especially I here wish to express my gratitude and admiration for his great devotion and patient perseverance.

I may also mention my indebtedness to Mr. C. van Werkhoven, who executed plates 1–6 with such admirable skill, thereby overcoming numerous difficulties with great ability. It was he too who put the finishing touches to several figures in the text. Most of the latter, however, I owe to Dr. R. de Wit, to whom I wish to express my sincere appreciation for his valuable aid.

And last but not least, I wish to extend my thanks to Mr. W. Nyhoff, who in spite of the present circumstances, has not hesitated to publish this book with the excellent care which so characterizes him.

Wassenaar, October 12th 1941.

Chapter I

SPACE AND TIME

"Our conception of the structure of the Universe bears all the marks of a transitory structure. Our theories are decidedly in a state of continuous and just now very rapid evolution".

(W. de Sitter)

Introduction

Geology, the science of the history of the Earth and Life, reaches back into the infinitely remote ages and depths of the Universe and extends its speculations to the origin and meaning of all organisms and anorganic matter.

One generation after another has attempted to unravel the problems of the continents and oceans, or to decipher the origin and evolution of Life, or even the mystery of Man himself, who never rests in his unceasing quest for knowledge.

The historical succession of phenomena, their correlation and meaning, form the most attractive and interesting feature of geology. This applies not only to the major outlines of the development of our globe and the evolution of life, but also to any other minor geological problem. Thus the geologist who is mapping a region keeps careful note of every detail of the rocks and strata. Nevertheless, his object is not merely to determine whether granite, limestone, schists or other rocks occur within the area, nor will the presence of folds, overthrusts or other tectonic phenomena satisfy his curiosity. What he wants to know is the sequence of events through space and time, i.e. the geological history of that particular region up to the present day. The only reward for his painstaking efforts will perhaps be what Termier so enthusiastically described as "la joie de connaître".

The geologist might be compared to the historian. The historian will make a careful study of any parchment that happens to fall into his hands and will reconstruct the past with the aid of its data if he finds them to be complete. On the other hand, he will be sure to point to the gaps in his knowledge if he finds them to be incomplete and realizes that he cannot arrive at the truth with absolute objectivity. In such circumstances he may possibly revert to other methods in an attempt to overbridge the missing parts, and will make use of temporary hypothetical constructions for lack of solid facts. It often happens that the historian, who supplies from his own imagination the missing lines of his scientific prose, allows himself to be carried away by an unbridled poetical inspiration and soars to giddy heights. This is inevitable, yet he should never forget that hypotheses constitute a necessary evil and should be discarded as soon as contradictory facts come to light.

We may expect to find a similar "geopoetical" aspect in many a geological treatise, in addition to the normal geological prose. However, authors should always keep their theories strictly separated from descriptions and conclusions of a more rigorously documented kind.

We repeat it: geology is a historical science. The history of the earth is a most absorbing one and its unknown elements — many of which will perhaps elude us forever — challenge us. The beginning of this history brings us into contact with Astronomy, particularly with Cosmogony, the science of the origin of the universe. Immeasurable space and time encompass us. "As Rama looks out upon the Ocean, its limits mingling and uniting with heaven on the horizon, and as he ponders whether a path might not be built into the Immeasurable, so we look over the Ocean of time, but nowhere do we see signs of a shore". These lines appear towards the end of the second part of Suess' masterly work *Das Antlitz der Erde*.

Science progresses with steady strides. New facts come to light and new ideas are born every day, and our conception of the structure of the universe changes accordingly. Our views have constantly to be revised and readjusted, but occasionally new aspects of far-reaching consequence shed such an unexpectedly different light on existing problems that its effects might be compared to those of a revolution. All that had hitherto been sacro-sanct crumbles to the ground, hardly anything is left untouched.

The late American geologist Barrell, whose death, alas, came so prematurely, wrote in one of his brilliant articles: "The scheme of the Universe is more profound and the unknown is a little nearer than it was recently thought to be. But such has been the progress of knowledge since man, in the days before the advent of science, naively regarded the earth, his home, as the center of the universe and the heavenly bodies as lights in a nearby firmament, created a few thousand years previously especially for his benefit".

In the light of the above, it would be advisable to begin with a brief outline of some of the features of the modern conception of space and time.

The universe, the solar-system and the earth

We will begin with a short summary of cosmic dimensions. These may help us to obtain a better idea of the earth's humble place in the universe. We will then immediately pass on to a discussion of the origin of the earth.

The earth's volume is more than a thousand times less than that of Jupiter, and the latter's volume is in turn a thousand times less than that of the sun (fig. 1). The sun looks like a small star when compared to a giant of the type of Antares. Sixty million suns could fill Antares' space, but this giant is surpassed several times by the super-giant Epsilon Aurigae, which has a diameter 3,000 times that of the sun. The sun represents only a small element of a spiral nebula, the so-called galaxy, composed of about ten to a hundred thousand milliard stars. A great many examples of this type of "nebulae" are known and their diameters vary between 1,000 and 100,000 light-years.

The planets of our solar-system lie extremely far apart, but the galaxy appears to be even more thinly populated with stars than the solar-system with planets. A few comparisons from Jeans may illustrate some of these cosmic dimensions. Five apples, placed on our five continents — one on Europe, another on Asia, etc. would provide us with a scale model of the stars' dimensions and the intervening distances. Supposing that — à la Jules Verne — we were to let ourselves be fired from the earth in a rocket, travelling at a rate of 5,000 miles an hour, it would take us two days to reach the moon. If we were to travel through the sun at the same speed, our journey would take us a week, and nine years would be required to pass through Antares. Finally it would take us no less than five thousand million years to travel in this same rocket through a spiral galaxy such as the Andromeda nebula (fig. 2, B).

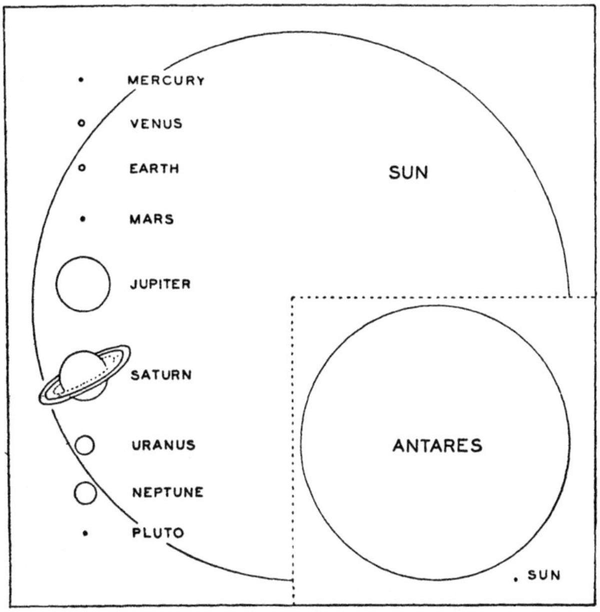

Fig. 1. The dimensions of the planets as compared to the disc of the Sun, and the Sun as compared to Antares.

Spiral galaxies are distributed less sparsely through space than the stars trough a galaxy. If Amsterdam — to quote an example from de Sitter — were to represent the extent of our galaxy its next-door neighbors would be the Hague and Utrecht. Some 800,000 light-years separate us from the nearest spiral nebula. In other words, if viewed from this region to day the earth would present a somewhat unwonted aspect for Man would just be beginning to appear. The spiral nebulae are distributed fairly evenly through space. The most distant ones — representing the very extremities of that part of the universe which is known to us at present — are separated from us by a thousand million light-years. This means that if it were at present possible to scan the earth trough a super-telescope from one of these extremely distant parts, we should find that we were looking at the first and most primitive terrestrial organisms in our history, swimming around in precambrian seas, and we should have to wait 700 million years to detect the first signs of life on the continents — that is, assuming that the distance remains unchanged during this lapse of time. These cart-wheel shaped galaxies have a central hub (fig. 2, B). Thirty thousand light-years separate the sun from the center of its sidereal system (fig. 2, A) and this galaxy rotates in approximately 200 million years according to observations by Oort, Lindblad and Plaskett.

One of the most sensational astronomical discoveries of the twentieth century was that all the star systems appear to recede from our galactic system, as well as from one another. Everyone knows that the pitch of a locomotive's whistle will drop as the engine rushes past and vanishes into the background. The same phenomenon is observed when a source of light recedes from us at high speed. It manifests itself to the astronomer as a red-shift of the star's spectral lines. This enables him to calculate the velocity of the stars' movement relative to the sun from the amount of the red-shift. Observations have disclosed the remarkable fact that the velocities of the recession of the spiral systems from our own galaxy and from one another, increase proportionately with the distance between the receding systems. By reversing the picture and imagining the galaxies to travel towards instead of away from one another, we are able to conclude that a tremendous quantity of matter was packed into a considerably smaller volume of space some 2,000 to 3,000 million years ago. The fact that all the galaxies move away from us, does not mean that we are remaining stationary. We should think in this respect of bits of straw which, while floating in a swiftly-flowing river, move away from one another as the river broadens. The statement that the universe is expanding (which was first alluded to by W. de Sitter in 1917 on theoretical grounds) is one of several possible interpretations of a mathematical equation. One possible solution is that the universe had once shrunk considerably in the past and that since then it has been continually expanding. Another is that the universe is subjected to alternating contraction and expansion, a pulsating movement. The last expansion must have begun at any rate some 2,000 million years ago, and this moment must have been one of major importance in the history of all cosmic matter. This question will be dealt with again presently, but it is obvious that our con-

Fig. 2. Above, our galactic system (after Easton). S indicates the solar-system. Below, the Andromeda nebula, a spiral galaxy seen from aside.

ceptions as to what exactly must have happened at that moment are, to say the least, of a very uncertain nature. The following reflections appear in a publication by de Sitter [1]: "It cannot be said whether the galaxies were already in existence before the catastrophe, or whether they originated from this turbulence as it may be that the stars were distributed more evenly through space previously. We know for certain, however, that all eventual deviations from absolute homogeneity existing at that time must have been strongly augmented by the terrific commotion". It would take us too far to mention the many interesting theories on the ultimate limit of the age of matter, which astronomers estimate to amount to between 5 and 10 billion years, or to deal with the absorbing results obtained as regards curved and finite space, and the correlation of matter and radiation, space and time.

The above quotation from de Sitter shows that the problem of the origin of the galaxies is far from being solved. This is equally true of the solar-system. There are still a great many conflicting views on this subject! Russell has reviewed the *embarras de choix* in a cleverly written book. One group of hypotheses assumes that there had been a close encounter between the sun and another star during some period in the past, and that both either approached one another very closely or collided. The chances are that such an event did actually take place 2,000 million years ago. A much debated hypothesis of Jeans and Jeffreys surmises that the sun thus entered the danger zone of gravitational pull of a bigger star, and that the sun's surface rose toward it in the shape of a conical point, which would have produced a narrow filament (fig. 3). The ejected material would then have condensed in separate cooling masses revolving around the sun, and would thus have led to the creation of the planets. Russell and Lyttleton are of the opinion that the sun might have been a binary star at that time, and that its smaller companion broke into fragments as the result of a collision with — or the near approach to — a passing star. These fragments would then have developed into the planets.

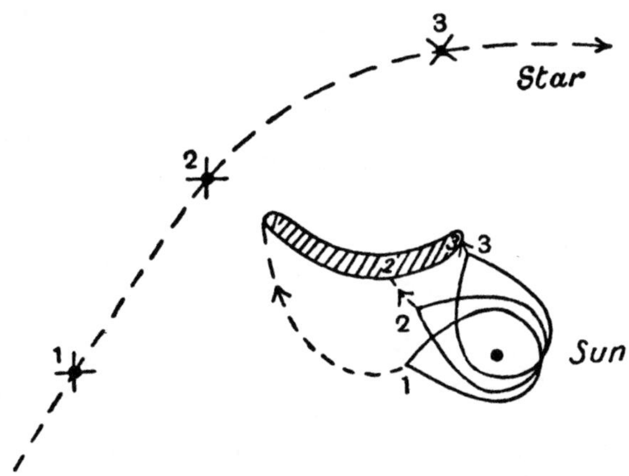

Fig. 3. Diagram of the changes in the form of the Sun and the formation of a boomerang-shaped filament of ejected matter during the passage of a star. The filament is shown in a position corresponding with position 3 of the star (After H. Jeffreys).

The birth of the moon

The solar-system's genesis presents many problems and is the subject of much conjecture. The same is true of the origin of the moon.

[1] De Sitter, 1934, p. 377 (translated).

The ratio of the volumes of the moon and the earth is 1 : 82. This is exceptionally high compared to that of Titan and Saturn (1 : 4700, the highest ratio found among the remaining satellites and planets). Besides, the moon's orbital momentum — which, in the case of other satellites, amounts to a mere fraction of the angular momentum of the accompanying planets — is five times that of the earth. This means that the system of the moon and the earth is more like a binary star than a miniature planetary system. From this it might be assumed that the moon originated as a separate body during the catastrophe which resulted in the formation of the planets. However, this hypothesis — if true — contains a few incomprehensible elements. It is difficult to see why two bodies, so unequal in size, should have originated during the catastrophe, and why these two bodies should have formed so close together. One thing is clear, however — the moon has slowly been retreating from the earth ever since its formation. The friction of the oceanic tides has gradually slowed down the earth's rotation and diminished its angular momentum. The result is that the moon's recession from the earth amounts to about five feet per century. Higher tides are observed as we reach back into the past; the days are found to be shorter, and the moon is seen to revolve nearer the earth.

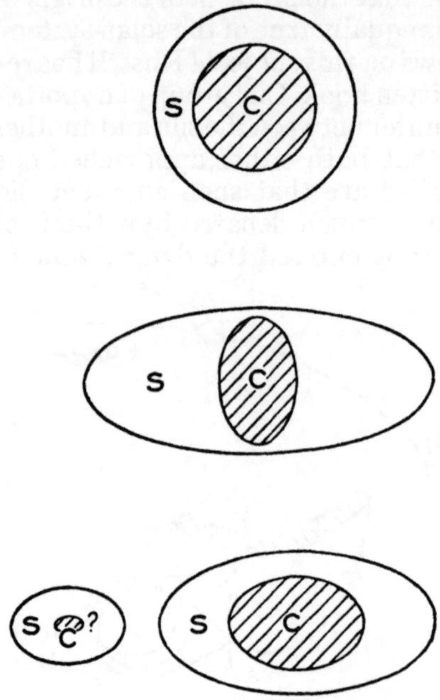

Fig. 4. Changes in the form of the earth during resonance, and the origin of the moon; c, nickel-iron core; s, intermediate and outer shells (After H. Jeffreys).

Of course, we sometimes wonder whether an entirely different hypothesis should not be preferred to the above one, and whether the moon and the earth had not at one time been united as a single body. This was the opinion of Sir George Darwin, who regarded the moon's disruption as a resonance effect, i.e. the result of a concurrence of the solar tide with the natural, free period of vibration of the earth's cooling fluid mass. The initial small amplitude of the tidal movement would continue to increase steadily according to this theory, and the globe would ultimately grow unstable and disintegrate when a certain amount was exceeded (fig. 4).

This hypothesis is based on the assumption that the moon had been severed from the earth during the earliest part of the latter's evolution. The liquid rocks of the earth were probably already more or less differentiated by that time, the heavier minerals sinking, and the lighter ones rising and accumulating in outer layers of the molten planet. In Goldschmidt's opinion a sulphide-oxyde melt probably separated around a central nickel-iron core, while the outer part of the mantle was formed by the relatively light material of a silicate melt (fig. 5).

A great part of the earth's outer shells clearly entered the formation of the moon. This fact not only explains why the moon's density is less than that of the earth, but also why its nickel-iron core is a comparatively smaller one or non-existent [1]), and why its external, silicate shell is so much thicker.

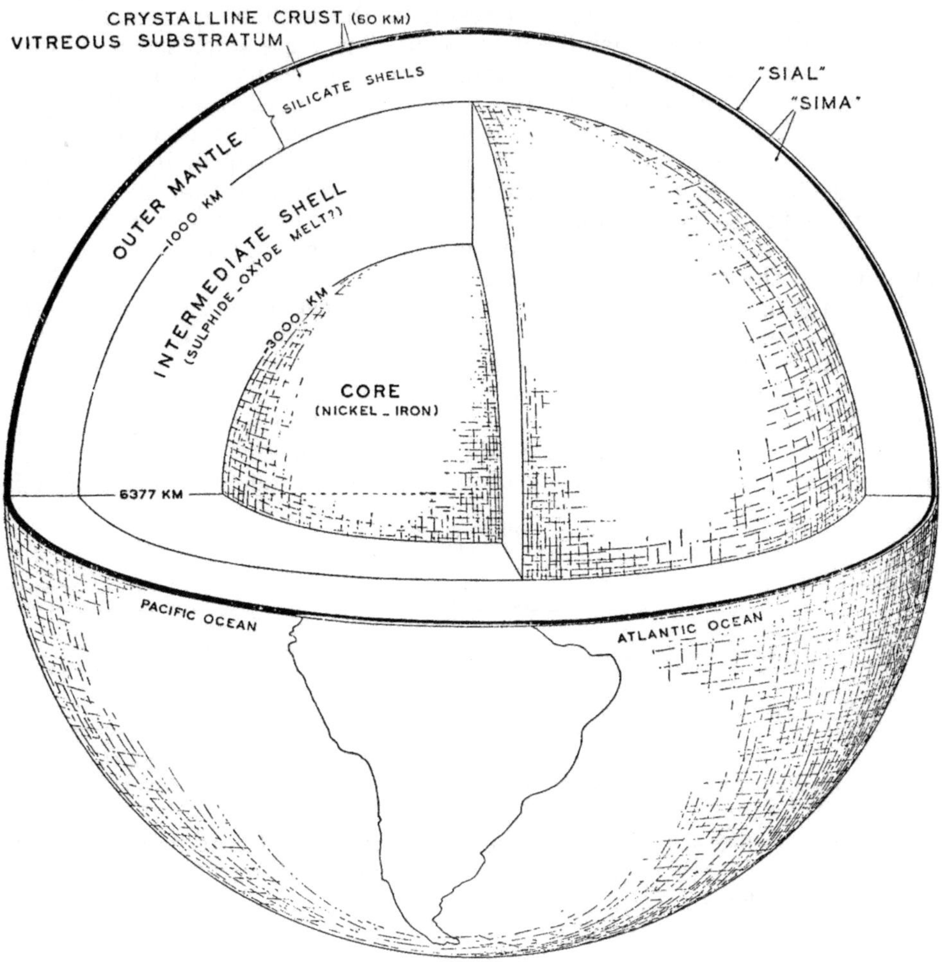

Fig. 5. Schematic and tentative illustration of the internal structure of the earth.

Fig. 5 gives a diagrammatic and tentative illustration of the earth's interior in its present state, showing the metallic core and enveloping shells,

1) Escher pointed out that the specific gravity of the moon's core would, on the earth's surface, amount to 3.89. This is more than the specific gravity of any rock on earth under a pressure of 1 atmosphere. "From this it would appear that some nickel-iron had passed from the earth to the moon. This must be understood in such a way, that no material of the core of the earth disappeared to the moon, but material from a ferro-sporadic shell" (Escher 1939, p.131).

which become increasingly siliceous or "acid" towards the outer part, and decrease in density. This schematic illustration has been made largely possible by the interpretation of seismic data. The outer, dark-coloured zone in fig. 5 represents the earth's crust, composed of some 40 to 80 km of crystalline rocks. This solid, elastic crust presumably rests on amorphous formations, i.e. rocks which, as a result of the high temperature at this depth, must be above their melting point. They are regarded as a fluid possessing a high viscosity. Daly speaks of a "vitreous substratum". The schematic cross-sections of the continents have been drawn as white lenses into the dark zone. Petrographical, volcanological and seismic evidence has lead to the assumption that the continents are built up of light siliceous material (55—70% SiO_2), viz. granitic "acid" rocks and sediments. This constitutes the so-called sial, a term introduced by Suess (the word sial is composed of the first syllable of silicon and aluminium, the most abundant elements of the continents). The sial rests upon less acid rocks (35—55% SiO_2), ranging from basalt to peridotite. These last materials build up the so-called "sima" — another petrographic term created by Suess and mnemonic of its most abundant elements, silicon and magnesium. It is believed to extend beneath the sialic continents. One of the ocean-floors — that of the Pacific — may probably be assumed to be composed of this simatic material in a crystalline state. As regards the Atlantic and the Indian Ocean, their bottoms are thought to consist of thin layers of sial,

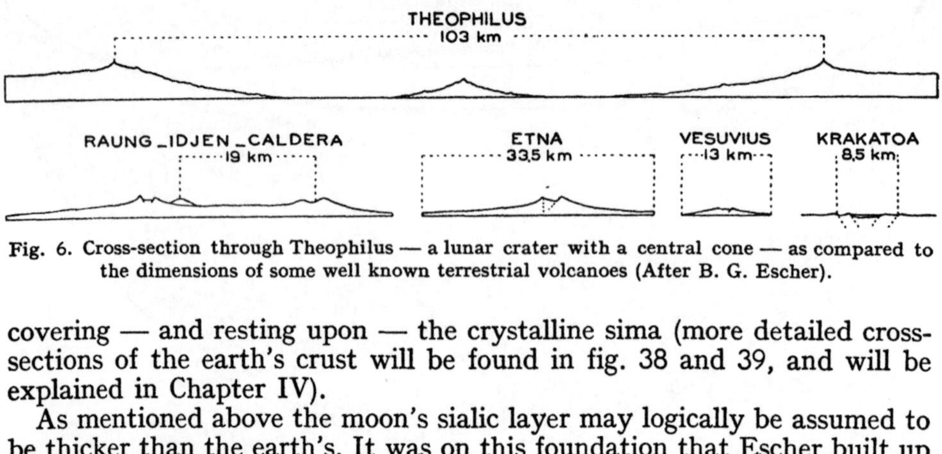

Fig. 6. Cross-section through Theophilus — a lunar crater with a central cone — as compared to the dimensions of some well known terrestrial volcanoes (After B. G. Escher).

covering — and resting upon — the crystalline sima (more detailed cross-sections of the earth's crust will be found in fig. 38 and 39, and will be explained in Chapter IV).

As mentioned above the moon's sialic layer may logically be assumed to be thicker than the earth's. It was on this foundation that Escher built up his interesting interpretation of the moon's morphology. A granitic magma abounding in gasses is known to bring about a much more explosive kind of volcanism on earth than a basic magma, and it follows that the discharge of gasses from the moon's very thick sialic shell caused a huge amount of volcanism, resulting in types of volcanic explosive vents (fig. 6) such as have never been equalled, either in abundance or magnitude, by similar terrestrial volcanoes [1]).

[1]) I endorse Escher's statements as regards the moon's volcanic morphology up to this point, but I believe that there are not enough facts with which to calculate the thickness of the lunar sial. We will return to this interesting point in Chapter VI, when discussing the problem of the origin of continents and ocean basins.

On the other hand several terrestrial phenomena may perhaps find an explanation in lunar influences.

We will therefore return to the primordial history of the moon and the earth in Chapters VII and IX, and especially in Chapter VI, when discussing the origin of the continents and oceans.

The age of the earth and the universe

Similar results as regards the age of the earth were obtained by investigators in different branches of science, and show that 2,000 to 3,000 million years have elapsed since its creation. These results were arrived at by petrographers and physicists in collaboration with geologists and also by astronomers. These figures are especially impressive in as much as many of these scientists arrived at them independently and their importance is augmented by the fact that — apart from the earth — many other bodies, such as the moon, the planets, the entire solar-system and the spiral galaxies, received the impulse to perform their present movements at the same moment. This moment seems to have formed the last critical date in the universe which since then has developed gradually into what it is like at present.

By means of no less than seven independent methods have scientists arrived at an estimate of the age of the universe since the last great catastrophe.

Chemists have shown that uranium disintegrates into radium and helium, and that radium will in turn disintegrate into lead and helium. Another important factor is that the radium lead which accumulates in this manner has a different atomic weight (Pb. 206) to ordinary lead (Pb. 207) or thorium lead (Pb. 208). An analysis can show us a rock's contents of lead generated from the decay of radium (we will not examine the problems affecting the determination and analysis of the various isotopes of lead), and as the ratio of radium lead accumulating from a given quantity of uranium per unit of time is known, the age of any rock in nature containing uranium minerals can be ascertained.

A second method is that based on the constant ratio observed between the amount of expelled helium atoms and the initial uranium. Helium is a gas, however, and can escape as such from a rock. The values arrived at by this method will therefore probably be too low. In other words, the ages of rocks obtained by the helium method will also be too low. Many ingenious improvements have been introduced since then to remedy this deficiency in an attempt to obtain useful results even from this method. The ages of a great many rocks have been determined so far, and their analyses showed that the ages of those rocks which could indeed be said to be older because of the geological position in which they were found, were invariably higher than others. The oldest rock ever analysed comes from Manitoba, and is 1,750 million years old (see fig. 7). The area in which it originated is known as one of the earth's most ancient structures, but many events preceded the formation even of this structure. A comparison between this region and other more recent and better-known areas makes it possible to calculate that the earth's

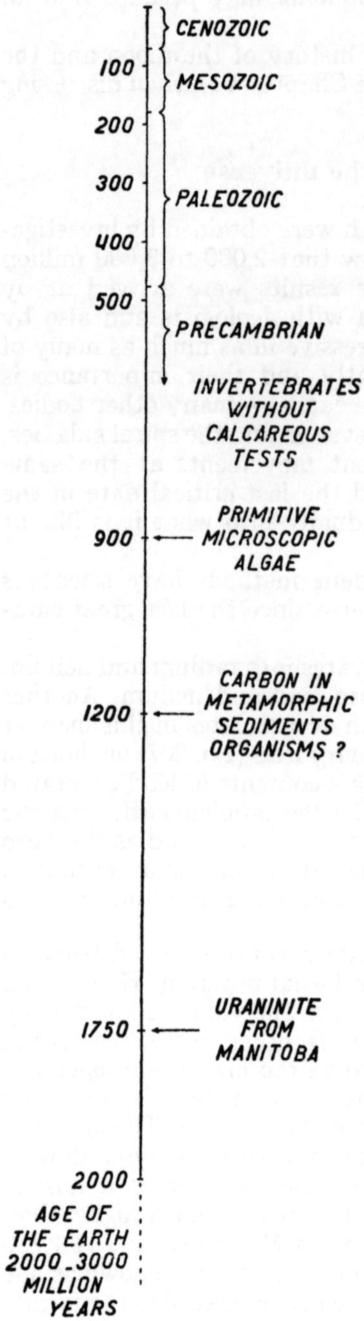

Fig.7. Graphic representation of the age of the earth.

age amounts to at least 1,750 plus approximately 300 million years, or some 2,000 million years.

This figure has not only been arrived at by geologists. It was mentioned above that a similar result was deduced from the recession of the spiral galaxies.

The spiral structure, too, may serve as a basis for further conclusions. Eddington showed that a galaxy could not possibly be preserved in a steady state for more than 10^9 years without considerable collapse or dispersal, and to this Bok added that "many spiral galaxies exhibit two well developed spiral arms and it is probable that such a structure cannot persist for much longer than ten to fifteen revolutions of the nucleus." As the galactic system revolves once in every 200 million years, its age might likewise be concluded not to exceed 2,000 to 3,000 million years.

These coincidences between astronomy and geology are truly remarkable, but we may add at the same time one more important argument. We know for certain that some of the nickel-iron meteorites which have dropped onto our planet can be said to have originated beyond the solar-system. A minute amount of radio-active minerals has been detected in some of these bodies, and their helium ratios have been determined. The eagerly awaited results of the investigation of these fragments of cosmic matter produced various figures, none of which exceeded 2,000 to 3,000 million years, however.

Another method is based on the movement of the moon. It was seen above that this body is steadily receding from the earth. If the process were reversed, the moon would be seen to be gradually approaching the earth. Though admitting that it is impossible to tell at what moment the moon has been severed from the earth, in as much as this process is itself purely hypothetical, Jeffreys estimated that the age of the moon was probably less than 4,000 million years. This figure is again remarkably near that of the 2,000 to 3,000 million years estimate mentioned above several times.

Lastly, Jeffreys calculated from the eccentricity of the planet Mercury that the age of the solar-system was probably nearer 1,000 than 10,000 million years.

This concludes our summary of the various methods which make it possible to calculate a minimum value of the age of the earth and the universe. It is obvious that these methods contain converging evidence of great importance.

A few methods are dealt with more extensively by Holmes in *The Age of the Earth*. We are endebted to the same publication for the data on the absolute age of terrestrial rocks as indicated in fig. 7 and Table II. These figures are extremely important, for they give us an idea of the duration of the different geological eras and periods. The latest investigations by Evans and his collaborators reveal that the helium ages of many geological horizons are to be lowered, but even their accurate methods of investigation leave the total span of geological time unreduced.

The history of the earth's crust

We usually divide the history of the continents into two distinct parts. The first deals with the remotest ages and covers the extremely long Precambrian era, of which very little is known. The second deals with the succeeding ages, from the Cambrian up to the present day. Much more is known of the earth's history since the Cambrian than of the preceding periods, for, generally speaking, its historical records have remained in a far better state of preservation than that of the earlier times. Historical geology has often been compared to a book, a torn and tattered old manuscript the pages of which lie scattered far and wide.The geologist's task is to hunt up the missing pages and replace them in their original order. The sediment layers can be said to represent the pages of this geological history and the fossils the writing. It should be noted that fossils occur very rarely in the Precambrian and are almost entirely absent in the older parts of this era. This explains why so little is known of precambrian history. Its pages can only be replaced in the correct sequence with the greatest difficulty (if at all), and the writing upon most of them has faded like that of an old palimpsest.

The principal events in the history of Life are listed in Table II, next to the time-scale. The Pleistocene and Holocene, encompassing the whole history of mankind from the earliest days up to our own, represent less than the thickness of the top line in fig. 7. Should our knowledge of the earth's history be communicated to a student in correct chronological proportion in the course of 50 to 60 lecture-hours, a bare two minutes would have to suffice for a discussion of the entire history of the pleistocene ice-ages. A comparison between the earth and various cosmic dimensions made us realize our planet's modest size. The preceding statement, however, shows that the history of the earth is at least 2,000 times as long as the whole history of *Homo sapiens*.

One of the most remarkable features of that history is the ceaseless flow of alterations. Mountains, seas, glaciers and deserts all seem immutable compared to the ephemeral span of human life. However, nothing is constant in geological time. Most visitors to the small though interesting town of Le Puy, in the south of France (fig. 8), come with the sole object of admiring its beautiful old cathedral and to gaze upon the enormous bronze statue of the Virgin. They stare in silent wonder at the steep rocks, each of which is crowned with a church or statue, but few realize that this strange scenery is merely due to the fact that the hard filling of a volcanic vent and some remnants of former lava flows are resisting erosion a little longer than the largely eroded tuffs and strata.

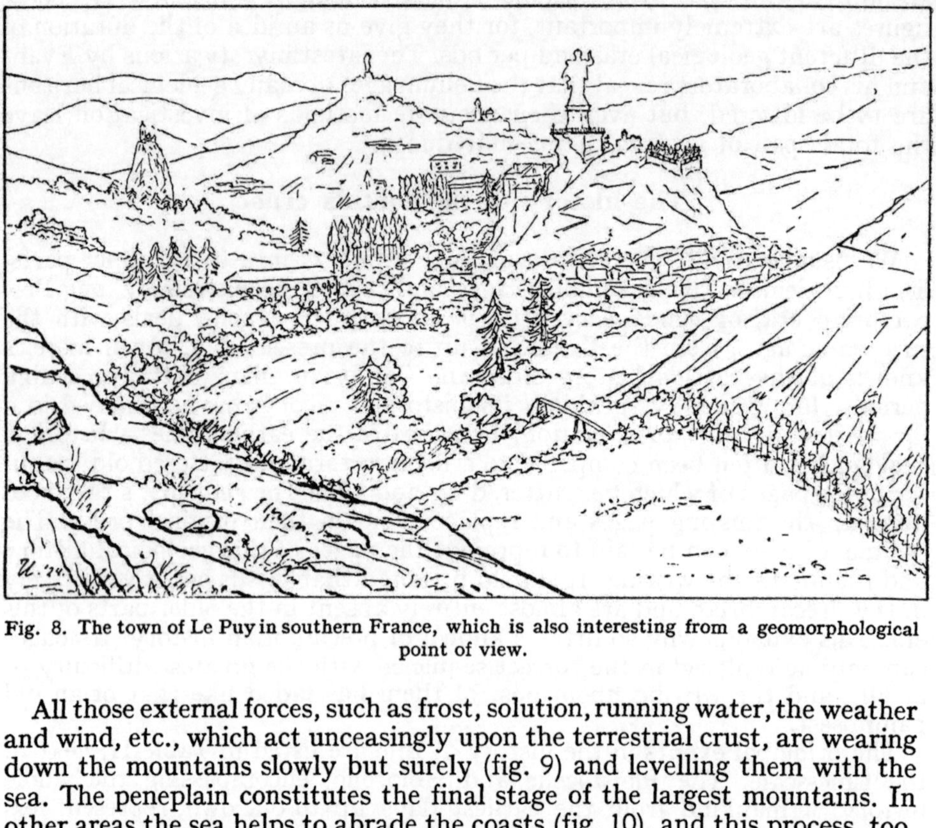

Fig. 8. The town of Le Puy in southern France, which is also interesting from a geomorphological point of view.

All those external forces, such as frost, solution, running water, the weather and wind, etc., which act unceasingly upon the terrestrial crust, are wearing down the mountains slowly but surely (fig. 9) and levelling them with the sea. The peneplain constitutes the final stage of the largest mountains. In other areas the sea helps to abrade the coasts (fig. 10), and this process, too, has a levelling effect. Neptune invades the land to an ever-increasing extent, and the débris of the organisms which remain behind enable us to determine the relative ages of the various deposits. An illustration of one of these marine strata, which was deposited during the Cambrian in a sea invading the land over Torridon sandstone of late-precambrian origin, will be found in fig. 11.

This erosive action, or denudation, is represented schematically in blocks 1 and 2 of fig. 12. Only a few of the remnants of this erosive action, or "monadnocks", as they are called, protrude from the peneplain in block 2 as evidence of the former relief. These two blocks show more, however. Rivers transport detritus from the mountains to the sea, where it settles in large quantities. The land represents an area of erosion, but the sea is the most suitable place for the accumulation of great quantities of erosion products. One layer slowly accumulates on top of another, and this leads to a thick sequence of so-called sediments. Block 1 shows that sedimentation takes place in a trough-shaped depression of the earth's surface. The coarsest products remain near the coast, the finer material is carried further away and settles in thin layers at a considerable distance from the shore. The coastline extends slowly but surely into the sea, and the sequence of sediments steadily increases. Block 2 shows that the floor of the basin of sedi-

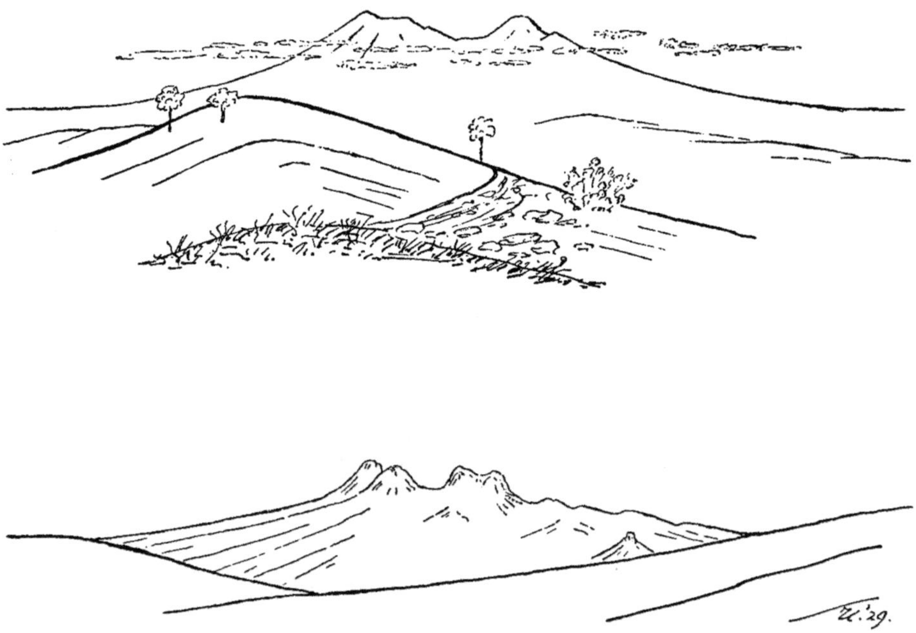

Fig. 9. The active Gedeh-Panggerango volcano (above) and the much eroded volcanic ruin near Plered (below), as seen from mount Bengkung, West of Bandung, Java.

mentation also subsides. Without such subsidence the thickness of the strata would never be able to exceed the original depth of the sea at this point. The accumulation of sediments, however, increases proportionately with the floor's subsidence. A thickness of many thousands of meters (sometimes as much as 15,000 meters or more) is not an unusual occurrence.

Everyone probably knows that not only are the sediments which settle near the coast unlike the strata deposited in the deep-sea, but that these separate environments also contain different kinds of organisms. Various factors such as light, food, pressure, temperature, salinity, the movement of the water and a host of other influences all affect the zonal distribution of organisms in the sea. An idea of the conditions under which these strata were originally formed can be obtained partly from their special lithological character but principally from the remains of former organisms. This total aspect of the sediments is known as the facies in geological terminology (fig. 13). The facies of sediments deposited in a shallow sea in the vicinity of the coast (littoral,) or in a few hundred meters of water (neritic sediments) are wholly unlike those which settled further away from the coast in

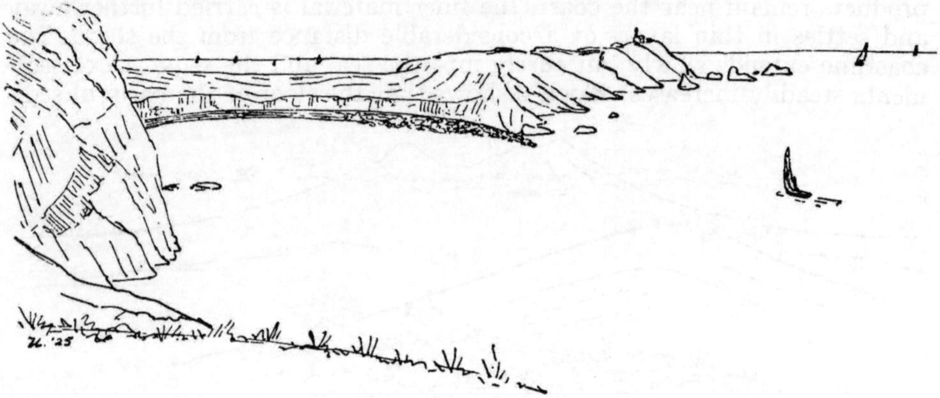

Fig. 10 The erosive action of the sea, near Camaret, Crozon-peninsula, in the neighborhood of Brest. The lower central part consists of precambrian schists; to the left and right lower-ordovician sandstones are exposed, forming an anticline.

approximately 1,000 meters of water (the so-called bathyal sediments) or at an even greater depth (abyssal sediments).

Littoral or neritic sediments can not form in more than a few hundred meters of water, but we sometimes come across a sequence of neritic sediments

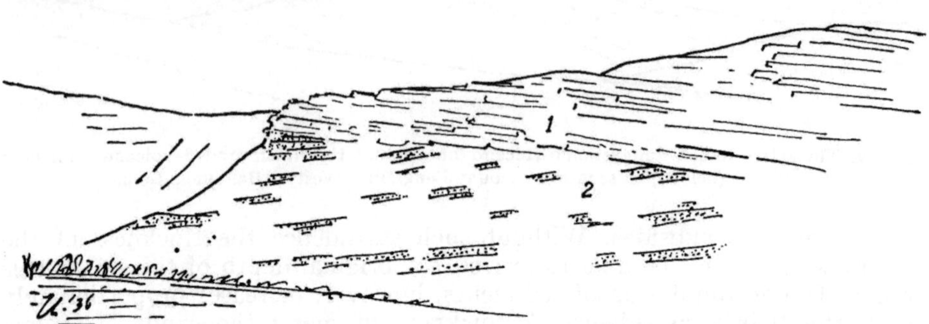

Fig. 11. Transgression of Cambrian quartzites (1) on precambrian Torridon sandstones (2). Quinac near Inchnadampff, in the northwestern Highlands of Scotland. (Compare with fig. 17).

totalling a few thousand meters in thickness, and it is clear that this must have been brought about by an accumulation of successive deposits during some slow subsidence of the sea floor. The name geosyncline has been given to these subsiding furrows of the terrestrial crust, into which sediments settled up to abnormally thick sequences. A schematic view of a geosyncline's cross-section will be found in block 2 (fig. 12). The downward movement of these formations must be attributed to the influence of deep-seated forces, and the same forces will at a given moment put a stop to the subsidence. A growing compression of the crust then causes the geosyncline to be thrown into spasms and its contents to be crumpled and folded. Block 3 depicts a more advanced stage. The whole zone will then tend to rise, and from the elongated belt emerges a mountain range which rises upwards very

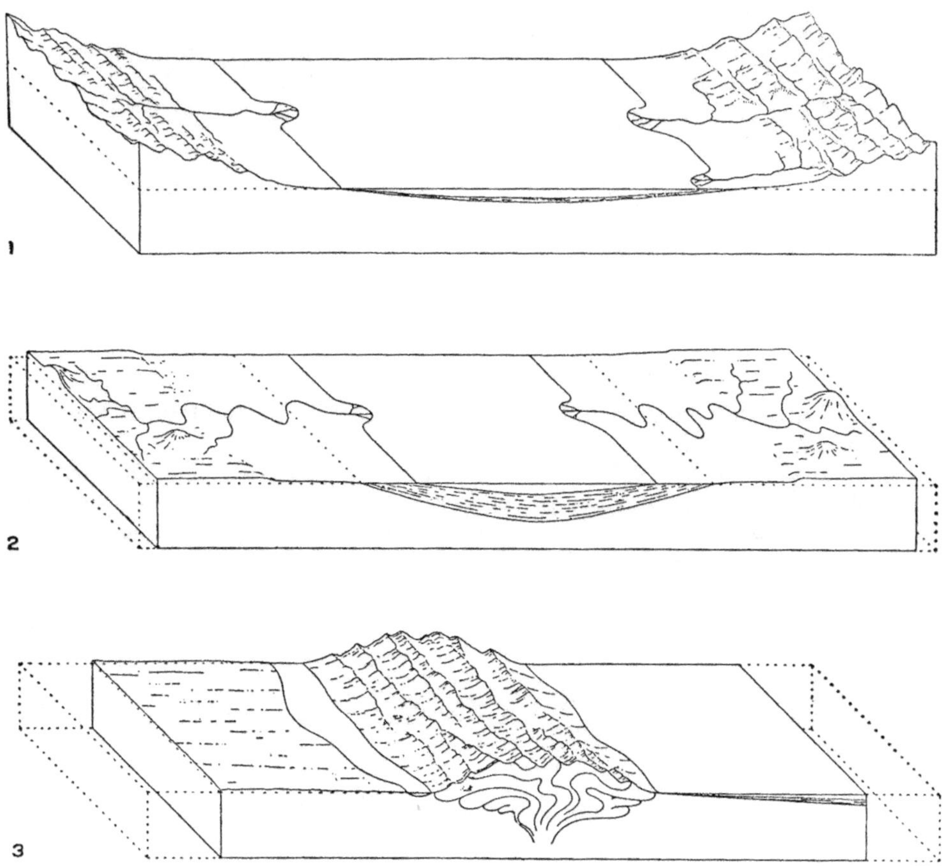

Fig. 12. Diagrammatic representation of a geological cycle.

slowly, though with unremitting regularity, frequently to a height of several thousands of meters.

The landscape reproduced in fig. 14 borders on the caledonian geosyncline in the north-western part of the Scottish Highlands. To the left of it are seen cambrian strata (these were deposited normally on an older precambrian basement, consisting in this case of Lewisian gneiss). To the right, however, a large "nappe" of Lewisian gneiss may clearly be seen to be overthrusting toward the left, i.e. over the edge of the geosyncline.

The external forces embark upon their destructive action as soon as the folded chain emerges. A new area of subsidence and sedimentation is formed elsewhere (this is indicated schematically towards the right of block 3, fig. 12); a new geological cycle has opened. The vertical proportions

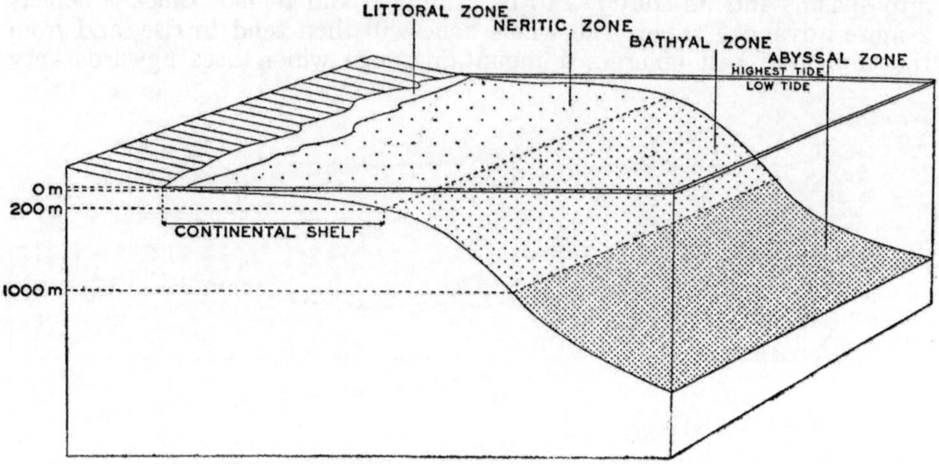

Fig. 13. Schematic blockdiagram of different facies types.

in fig. 12 are exaggerated, but these rhythms will be dealt with again in Chapter IV (fig. 43).

A natural section such as that in fig. 15 provides the key to the reconstruction of a geological cycle. The strata in this consist of devonian Old Red sandstone resting unconformably on steeply dipping silurian

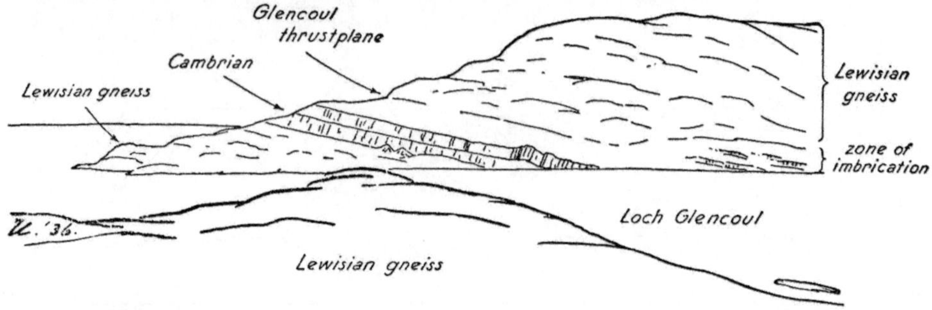

Fig. 14. Glencoul overthrust. Assynt Mountains, North-West Scotland (Compare with fig. 17).

sandstone. A plane of unconformity has a long history. The marine limestones in fig. 15 were deposited normally, i.e. just about horizontally, during the Silurian, and were subsequently steeply folded and elevated above sea-level. Old Red sandstone later settled on top of these eroded and peneplained marine strata. The latter's period of folding corresponds with that in which a geosyncline was compressed over large areas in Scandinavia, Scotland and Ireland. This last zone was later raised as a mountain-chain (Plate 1 indicates this belt with green lines). The actual Old Red sandstone was supplied by detritus from these chains, and was later likewise worn away by erosion in extensive areas. Trough-shaped depressions, elongated in the trend of the Caledonian mountains, formed here and there (fig. 28), and the Old Red only escaped erosion in these depressions. Some of the

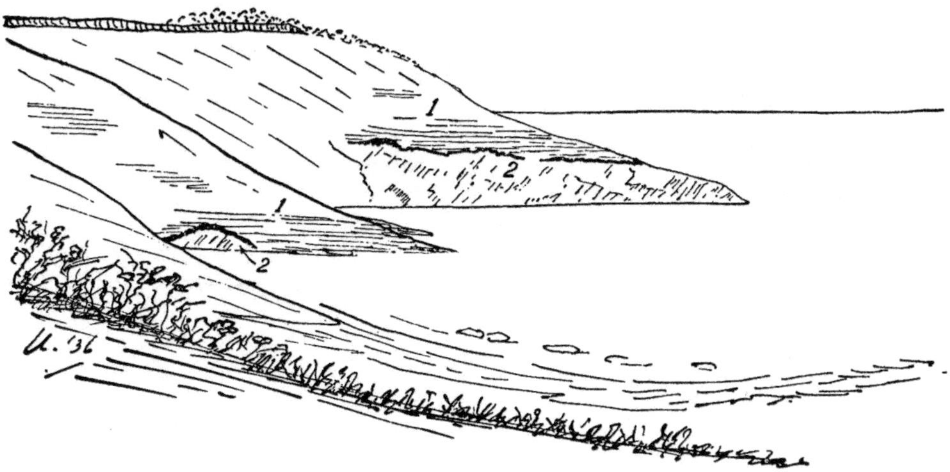

Fig. 15. Old Red (1) resting unconformably on steeply dipping silurian limestones (2). Siccar point near Dunbar, Scotland. (Compare with fig. 17).

troughs are bounded by faults (fig. 17), in which we once more find the trend of the old caledonian geosyncline. These faults are observed along the surface over a considerable distance (fig. 16) Besides, the relation between the scenery and structural history of an area is finely and instructively illustrated by the whole of Scotland's morphology. A diagram of some of Scotland's most salient geological and "geomorphological" features will be found in fig. 17.

In a few cases geosynclinal subsidence and folding are known to have occurred several times in the same region. One of these complicated areas is depicted in fig. 18. In the Ardennes, namely — near Fépin, along the banks of the river Meuse — the lower paleozoic deposits (Revinian) of an ancient caledonian geosyncline were strongly folded and covered unconformably by lower devonian (Gedinnian) strata, beginning with a basal conglomerate. This younger sequence of the so-called variscian geosyncline was later refolded, and the plane of unconformity was also strongly undulated.

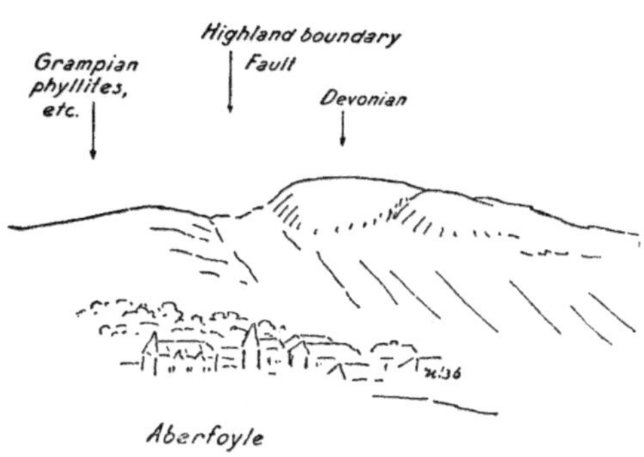

Fig. 16. The Highland boundary fault near Aberfoyle. (Compare with fig. 17).

Fig. 17. Geological diagram of Scotland, showing the localities of Fig. 11, 14, 15, 16 and 19.

The object of the preceding remarks, especially of the rough sketches in fig. 12, is to give an idea of the constant transformations that are altering the face of the earth. It should at once be added, however, that one of the most spectacular features of these transformations is periodicity.

The geologist comes across periodicity in many of the pages which he is so arduously deciphering, — in the sequence of the strata, for instance, and their contents of former organisms (stratigraphy). He observes it elsewhere, in the pulse of the deep-seated forces that bring about subsidence first in one area and then in another, culminating in the folding of geosynclines; and again — in the intrusion of liquid melts or "magma" rising from some deeper part of the earth's interior; in the rhythmical invasion of the continents by epicontinental seas and the latter's subsequent retreat (transgression and regression); he reconstructs the pulsation of the climates and the rhythmical evolution of Life.

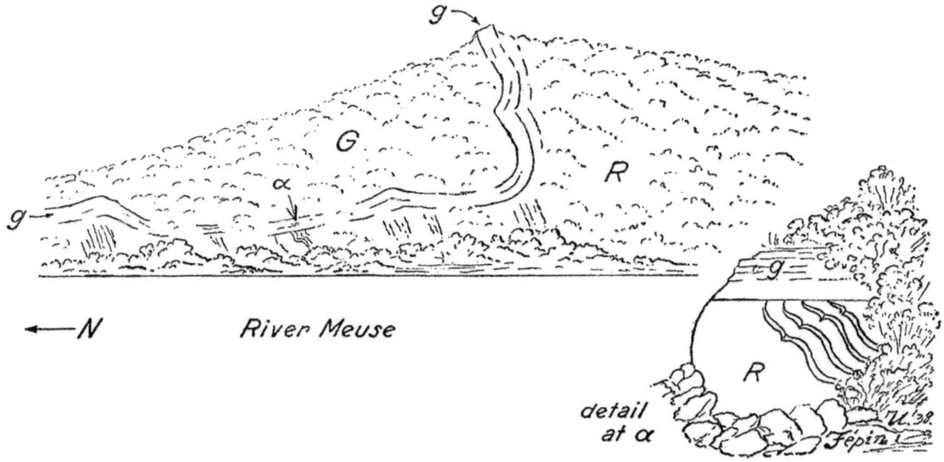

Fig. 18. The folded unconformity near Fépin in the Ardennes, a result of repeated geosynclinal subsidence and folding. G, Lower Devonian with basal conglomerate (g); R, Revinian.

Barrell's outstanding publication of 1917, which every geologist ought to read and re-read, opens with similar thoughts: "Nature vibrates with rhythms, climatic and diastrophic, those finding stratigraphic expression ranging in period from the rapid oscillation of surface waters, recorded in ripple-marks, to those long deferred stirrings of the deep-imprisoned titans which have divided the earth's history into periods and eras. The flight of time is measured by the weaving of composite rhythms — day and night, calm and storm, summer and winter, birth and death — such as these are sensed in the brief life of man. But the career of the earth recedes into a remoteness against which these lesser cycles are as unavailing for the measurement of that abyss of time as would be for human history the beating of an insect's wing. We must seek out, then, the nature of those longer rhythms whose very existence was unknown until man by the light of science sought to understand the earth".

Periodicity is the fundamental conception upon which the following pages have been built. The sequence of geosynclines and folded mountain-chains will be analysed in Chapter II. Chapter III will discuss another conspicuous feature of continental structure — the basins. The deeper structure of the earth's crust will be dealt with in Chapter IV, which moreover, will attempt to show the relation between tectonic cycles and the closely associated magmatic phenomena. Chapters II —V deal chiefly with the history of the continents and the oscillations of the sea-level. The problems arising from the ocean-floors, will subsequently be examined in Chapter VI. Chapter VII discusses the intricate problems of abnormal climates (the ice-ages). Chapter VIII attempts to trace the presence of a certain rhythm in the evolution of Life, especially in the evolution of the flora, and Chapter IX ends with a brief review of all the preceding phenomena which form *the pulse of the earth.*

References

BARRELL, J. *The strength of the Earth's Crust* (Journ. of Geology, vols. 22 and 23, 1914 and 1915).

BARRELL, J. *Rhythms and the measurement of geologic time* (Bull. Geol. Soc. of America, vol. 28, 1917).

BOK, B. J. *Galactic dynamics and the cosmic time-scale* (The Observatory LIX, Nr. 142, 1936).

DALY, R. A. *Architecture of the Earth* (1938).

EDDINGTON, A. S. *The rotation of the galaxy* (Oxf. Univ. Press 1930).

EDDINGTON, A. S. *The expanding Universe* (Cambridge 1933).

ESCHER, B. C. *Moon and Earth* (Proceed. Kon. Acad. van Wetensch. Amsterdam, vol. 42, 1939).

ESCHER, B. G. *Het probleem der caldeira's en der maancircussen* (Geologie en Mijnbouw 1, 1940).

EVANS, R. D., GOODMAN, C., KEEVIL, N. B., LANE, A. C. AND URRY, W. D. *Intercalibration and comparison in two laboratories of measurements incident to the determination of the geological ages of rocks* (Physical Review 55, 1939).

GOLDSCHMIDT, V. M. *Geochemie* (Handwörterb. d. Naturwissensch. 2d ed. 1933).

GUTENBERG, B. *Physics of the Earth* (vol. VII. Internal constitution of the earth 1939).

HOLMES, A. *The age of the Earth* (Th. Nelson & Sons, London 1937).

JEANS, J. *Astronomy and Cosmogony* (Cambridge 1928).

JEANS, J. *Through space and time* (Cambridge Univ. Press 1934).

JEFFREYS, H. *The Earth. Its origin, history and physical constitution* (Soc. Ed. Cambridge Univ. Press 1929).

LYTLETON, R. A. *The origin of the solar system* (Monthly Not. R. Astron. Soc. 96, 1936).

PLASKETT, J. S. *The dimensions and structure of the galaxy* (Oxf. Univ. Press, 1935).

RUSSELL, H. N. *The Solar System and its origin* (Edit. Mac Millan Cy, 1935).

SITTER, W. DE, *Kosmos* (A course of six lectures on the development of our insight into the structure of the universe delivered for the Lowell Institute in Boston, in November 1931, Cambridge Mass, 1932).

SITTER, W. DE, *Het uitdijend heelal* (Werken v.h. Genootsch. v. Natuur- Genees- en Heelkunde. Amsterdam, 2e ser., deel 14, 1934).

TERMIER, P. *La Joie de Connaître* (1926).

UMBGROVE, J. H. F. *Periodicity in terrestrial processes* (Americ. Journal of Science 238, 1940).

CHAPTER II

MOUNTAIN-CHAINS

Introduction

> "The profound revolutions, marked by folding, magmatic invasion and regional metamorphism were relatively brief periods closing long eras marked by diastrophic quiet and low continental relief". (J. BARRELL)

The first critical analysis of the earth's structural history was made by Ed. Suess, who published a monumental book on this subject: *Das Antlitz der Erde*. Suess was not only the right man for such a task, but he also undertook it at the right moment. The amount of data had already become so voluminous by then that an attempt could be made, by grouping them systematically, to analyse the surface of the entire earth. It would be impossible for a single man to perform such a task to-day, since a judicial examination of the enormous amount of world-literature would require more than a life-time. However, a series of regional treatises makes it much easier for us, at present, to find our bearings in this varied mass of publications and a further asset is, that the study of them requires far less of our time. Such synoptic works make it possible to form an idea of the whole region in question, without entering into a critical examination of all orginal data and fundamental details. In this manner it is possible to obtain a survey of all that appears to be known at present of the structural history of the continents and to base an opinion on its general aspects. This method had already been adopted previously by Born but he grouped his pictures regionally, i.e. continent-wise. We, on the other hand, have chosen a purely chronological sequence. Nevertheless, our two methods agree fully in one respect, i.e. the plain facts are stated separately, without being influenced by any preassumed hypothesis, and these are left to speak for themselves as they accumulate. The data have lastly been combined into a synoptic picture (Pl. 5), serving at the same time as a chronological analysis of continental structural history. At the end of the chapter the reader will find a list of the literature consulted. A summary and discussion of the various data will appear in the Appendix.

The following historical analysis will begin with the Cambrian, for the ages of the complicated structures of the precambrian basement have only been determined in a few cases, and the study of their intricate patterns can therefore still be said to be in its infancy.

The intensive folding of the Ardennes had long been known to have occurred at a much earlier date than the tectonic activity of such mountains as the Alps, the Pyrenees and the Himalayas. The upper-paleozoic chains were called the Hercynian Mountains, or — as this first name has now become obsolete — the Variscian (or Variscan). It had also been realized for some time that the final, intensive folding in Norway and Scotland occurred even earlier. The mountains in these last two areas were called the Caledonides. Nevertheless, as more and more facts came to light it became clear that the Lower-Paleozoic or "Caledonian" belts had not all originated during the same period. Thus the Caledonides of North-America, to mention but a few examples, were folded towards the close of the Ordovician, while those in Scotland and Ireland underwent a similar process some 40 million years later, towards the end of the Silurian (Gotlandian). We speak of a Taconic epoch of compression in the first case, and of an Erian in the second (fig. 19). Several diverging periods of origin are also observed in variscian and

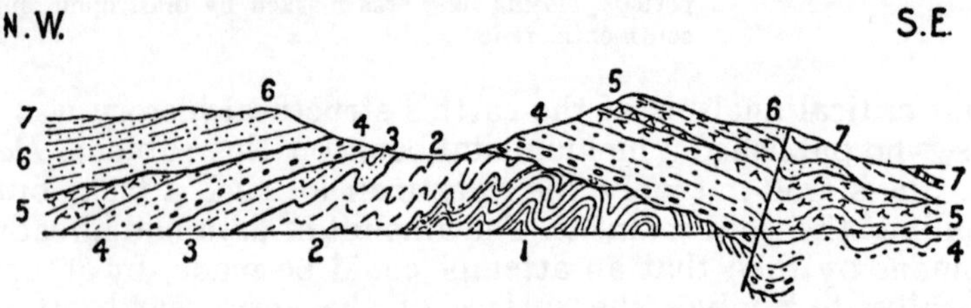

Fig. 19. Diagrammatic section across the Pentland Hills near Edinburgh (After W. H. Laurie), showing Erian and Mid-Devonian epochs of movement; 1 and 2 Silurian; 3 Downtonian; 4 and 5 Lower Old Red; 6 Upper Old Red; 7 Carboniferous. (Compare with fig. 17).

more recent mountains. A long list of epochs of compression has thus gradually been disclosed to us, the most important of which are submitted in Table II.

Another important point is that the epochs are distributed over the whole world. A comparison between said phases and their distribution (See Pl. 1—4) will at once render clear that one movement was of particular importance in one group of mountains, while in another chain the influence of a different diastrophic epoch was especially strong. It would not always be easy, however, to state concisely which epochs made themselves felt in Caledonian chains [1]), and which other ones in Variscian [2]), for Caledonian and Variscian epochs overlap. Moreover, the same applies to Variscian and younger convulsions.

Stille classified the post-Variscian Mountains as "Alpine" formations.

1) Though the oldest epoch in a so-called "Caledonian" belt is as a rule the Taconic, and the youngest the Erian or Mid-Devonian (fig. 19) far more ancient Cambrian phases seem to occur in Asia. The oldest movements in Australia (Cambrian and Ordovician) were likewise regarded by Andrews as Caledonian epochs, and the youngest epochs in this "Caledonian" sector were observed to occur towards the close of the Devonian (the Bretonic epoch appearing elsewhere as the oldest in Variscian folded chains).

2) For while the Variscian era of folding ends in most cases with a Pfalzian, and more often an even younger epoch "variscian" mountain-building in the Dobrutcha, the Falkland-Islands and the Cape Mountains appear to close with the Lower-Cimmerian phase, whose influence may be described as more or less evident in many cenozoic and mesozoic chains.

	Formations		Principal Epochs of compression	
	Pleistocene			
		. . . .	Wallachian (= Pasadenian)	
	Pliocene	Sicilian		
		Astian		
		Piacentian		
		. . . .	Rhodanic	
		Pontian		
		. . . .	Attic	
		Sarmatian		
Tertiary		Tortonian		
	Miocene	. . . .	Late Styrian	Alpine epochs
		Helvetian		
		. . . .	Early Styrian	
		Burdigalian		
		Aquitanian		
		. . . .	Savian	
	Oligocene	. . . .	Pyrenean	
	Eocene			
		. . . .	Laramide	
		Senonian		
	Upper Cretaceous	. . . .	Subhercynian	
		Emscherian		
		Turonian		
		Cenomanian		
		. . . .	Austrian (Oregonian)	
	Lower Cretaceous	Gault		
		Neocomian		
		. . . .	Late Cimmerian (Nevadian)	Mesozoic epochs
Mesozoic		Wealden		
	Jurassic	Tithonian		
		. . . .		
	Triassic	Rhetian		
		. . . .	Early Cimmerian (Palisade)	
		Norian		
		Karnian		
		Ladinian		
		Anisian		
		Skytian		
		. . . .	Pfalzian	
	Permian	Thuringian		
		Saxonian		
		. . . .	Saalian (Appalachian)	Variscian epochs
		Artinskian		
	Carboniferous	Stephanian		
		. . . .	Asturian (Arbuckle)	
		Westphalian		
		Namurian		
		. . . .	Sudetic (Wichita)	
		Lower Carboniferous		
		. . . .	Bretonic (Acadian)	
	Devonian	Famennian		
		. . . .		
		Frasnian		
		Givetian		
Paleozoic		. . . .	Mid-Devonian	
		Eifelian		
		Coblentzian		
		Gedinnian		
		. . . .	Erian	
	Silurian (= Gothlandian)	Downtonian (incl. Upper Ludlow)		Caledonian epochs
		. . . .	Ardennian	
		Lower Ludlow		
		Wenlock		
		Tarannon		
		Llandovery		
		. . . .	Taconic	
	Ordovician	. . . .	Ordovician	
		. . . .	Sardic	
	Cambrian	. . . .	Cambrian	
		. . . .	Late Precambrian	

I have divided them into two large groups, for the final phase is now known to have occurred during the Mesozoic — that is, before the major movements in the Alps — not only in the Sierra Nevada of North-America, but also in many other areas, e.g. in the eastern and south-eastern part of Asia. These belts have consequently been grouped separately (Pl. 3). The Cenozoic chains, beginning with the Laramide, will be found in Plate 4 (reference has also been made to mesozoic movements where these are known to occur in their history).

The different belts can thus be divided into four large groups: the early Paleozoic (Caledonides), late Paleozoic (Variscides), Mesozoic, and Cenozoic (Alpine *sensu stricto*). The structure of the continents will be reviewed accordingly in Plate 5 [1]).

The accompanying table lists the diverse epochs of compression. A more diagrammatic view of their place on the time-scale will be found in Table II.

We will now review the more general aspects of the data discussed in the Appendix [2]).

Epochs of compression

Our historical analysis of the continents begins some 500 million years ago, for too little is known of events prior to that time. The significance of such a statement will be realized if we pause to consider that the earth is thought to be approximately 2,000 to 3,000 million years old, for this means that we only begin to find sufficient evidence of the past at a time when 3/4 to 5/6 of the earth's structural history had already elapsed. In other words, only one fourth — or possibly even less — of this history can be said to be legible. So little is known of the preceding era — the Precambrian — that we prefer to leave it alone altogether. This Precambrian era, however, not only covers the greater part of time, but the precambrian areas also occupy the largest part of the continents. It should furthermore be noted that the continents, including probably large areas beyond them, had already passed through an extremely lengthy and complicated evolutionary stage before the Cambrian, and that hardly anything is known of this part of their development. Not only will a glance at the structural map in Plate 5 show that precambrian formations occupy extensive areas, but the precambrian basement is also known to crop out in almost all the larger Paleozoic, Mesozoic and Tertiary Mountains, proving clearly to what extent the younger part of the earth's history is engraved across the existing records of earlier ages.

1) These four groups can be subdivided into several others if classified according to the period of the belts' formation, i.e. the youngest epoch of compression appearing in their history. The Caledonian Mountains can thus be subdivided into the Taconic and late Caledonian; the Variscian into the Sudetic, Asturian and Saalian; the Mesozoic into the early-Cimmerian, late-Cimmerian and Subhercynian; the Alpine into the Laramide, Pyrenean, Savian, Styrian, Attic and Wallachian. These groups would obscure the main outlines if drawn into a single structural map with separate notations, and we have therefore only indicated the four principal groups in Plate 5. Other details will be found in Plates 1—4 and in the Appendix.

2) The maps do not show the tectonic character of the structures. Mountains with intensive overthrusts, such as the Alps, appear side by side with other less intensely folded mountains, e.g. the Rockies and chains of the Jura type. It will be obvious that the coloured lines only indicate the general direction of the strike in a schematic manner.

We often read of the accretion of the continents, meaning that younger geosynclinal belts and folded chains are arranged consecutively around a precambrian nucleus. In several cases these zones can be said to be younger the greater their distance from a nucleus (we will deal with this matter presently). There are indeed areas which demonstrate the outward displacement of zones of folding. It would not be right, however, to conclude from the above that a number of small blocks had existed in the Cambrian, which have grown larger and larger periodically, nor that the continents have consequently grown larger than at any previous date. On the contrary, it would be more plausible to assume that the continents had at one time been more extensive than they are at present. Not only have parts of the continents foundered below sea-level since Precambrian time, but they have even done so until quite recently, and their subsidence occasionally attained great depths! The present continents are mere fragments of one-time larger blocks. The same applies to the so-called "growth-zones". The Caledonides of Europe, North-America and Asia originally covered wider areas than their actual fragments do to-day. The latter are bounded and cut off by the frontal chains of Variscian and later mountains. The mesozoic geosynclines of Asia, North- and South-America still contain the vestiges of the former variscian basement upon which they were formed. Most of the Alpine chains of Europe and Asia are situated in a zone which had already been folded during Variscian time.

It cannot be denied, however, that there are cases in which the younger the age of the geosyncline, the greater the distance between the frontal chains and a precambrian nucleus. Here, too, a glance at Pl. 5 will be enough to show that this applies to Asia, Australia, and partly to other continents. Certain precambrian shields seem to form centres, around which originate a series of geosynclinal belts, spreading in ever-widening arcs. From Australia the caledonian, variscian, mesozoic and, finally, tertiary zones migrate east-and north-eastwards in a centrifugal sequence. An identical centrifugal displacement is observed around the Angara Shield in Asia, though with some complications. For the chains were compelled to bend around the Tarim and Ordos Massifs, including other precambrian nuclei, which had since long been rigid blocks, and analogous centrifugal waves spread from the Kara-Sea and Tschuktschen Massifs. The phenomenon of centrifugal migration can be discerned to some extent in Europe, viz. in a southerly direction. These examples seem to imply that certain centrifugal impulses radiate from specific precambrian areas. We might describe them as dynamic centers of tectonic activity (this is what Leuchs and Baily Willis called them). Nevertheless, there can be no question of a "law" in this case, for it should at once be noted that such dynamic waves have never emanated from other precambrian areas — neither from India, nor from Africa (with the possible exception of its extreme southern part), nor from the northern, western or southern part of Australia, nor from the eastern part of South-America. Leuchs therefore quite rightly described the block of India as a passive element in the structure of the Asiatic continent. It acted as an obstruction, against which variscian, and subsequent tertiary tectonic waves were shattered, and was responsible for the narrowing of the tertiary zone of folding and the particularly intensive overthrusting and elevation of the Himalayas.

However, it would be a mistake to suppose that the formation and migration of various zones of folding were due exclusively to the mysterious activity of these "dynamic centers". For before passing on to an examination of their remarkable sequence through space and time, another equally important circumstance should be noted, viz. the world-wide occurrence of epochs of compression. Data have frequently shown that simultaneous movements occurred on all continents. One phase was of course of greater importance in one place than another, manifesting itself more intensely in one area than elsewhere. But the growing amount of stratigraphic studies make it increasingly evident that the terrestrial crust was subjected to a periodically alternating increase and decrease of compression. This concurrence between epochs of folding is not only found in geosynclines, but also in other areas beyond them — in basins and regions where folding had already occurred at a previous date, and where less accentuated movements accompanied the folding of the geosyncline (shifting of blocks, faulting, sliding, and at times even folding). I feel there is overwhelming evidence that the movements are the expression of a common, world-wide active and deep-seated cause.

The term "world-wide" should not be interpreted to mean that all phases can be observed everywhere, or that these movements occur in all areas. On the contrary, the special constellation of the basement may cause two geosynclines to form side by side, one of which will be folded to an intense degree during a given epoch of compression, while the other will be hardly influenced by it. A striking illustration of this is found in the Cordilleras along the west-coast of North-America (fig. 20). The following are found side by side in this area: (1) the geosyncline of the Rocky Mountains, which was formed in the Precambrian and influenced by laramide folding; (2) the geosyncline of the Sierra Nevada (Triassic and Jurassic, nevadian folding); and (3) a geosyncline extending further west on a basement which was folded during the nevadian revolution: the geosyncline of the Coast Ranges (Cretaceous and Lower-Tertiary, miocene folding).

The same phenomenon can be observed in the paleozoic Wichita geosyncline of Central-America. The Wichitas and Criner Hills, both of which are situated in this belt, were influenced during the Wichita and the Arbuckle epoch, but the Ardmore basin in the immediate neighborhood of the geosyncline, including the Arbuckle chains, were only affected by the latter.

A certain similarity is apparent in the linear shift of specific epochs of folding in mountain-chains, e.g. in the Dobrutcha, Crimea, Caucasus; the Donetz Ranges, Ammodetic Mountains, and the Caledonides and Variscides on either side of the Atlantic Ocean.

These general aspects will help us with the further examination of the "dynamic centers". In which direction does that dynamic activity displace itself? The investigator would fain detect some regularity in its migration through space and time, some law, which at times we imagine can be formulated, though later it transpires that nature is more complicated than we supposed it to be. The sequence of the folded systems seems quite regular in Asia and the eastern part of Australia, but a far less simple state of affairs prevails in Europe. The Caledonides have been preserved along the north-western margin of the Baltic Shield, and perhaps originally bent around in a south-easterly direction. The Variscides, on the other hand, were not only partly formed in a caledonian basement (fig. 18), but the front of these folded chains intersects the trend of the Caledonides in Ireland and England (Pl. 5). This is also true of the younger folded chains in North-America. The latter intersect the trend of the Variscian Marathon-Ouachita Mountains, and these in turn cross the Wichita-Arbuckle geosyncline.

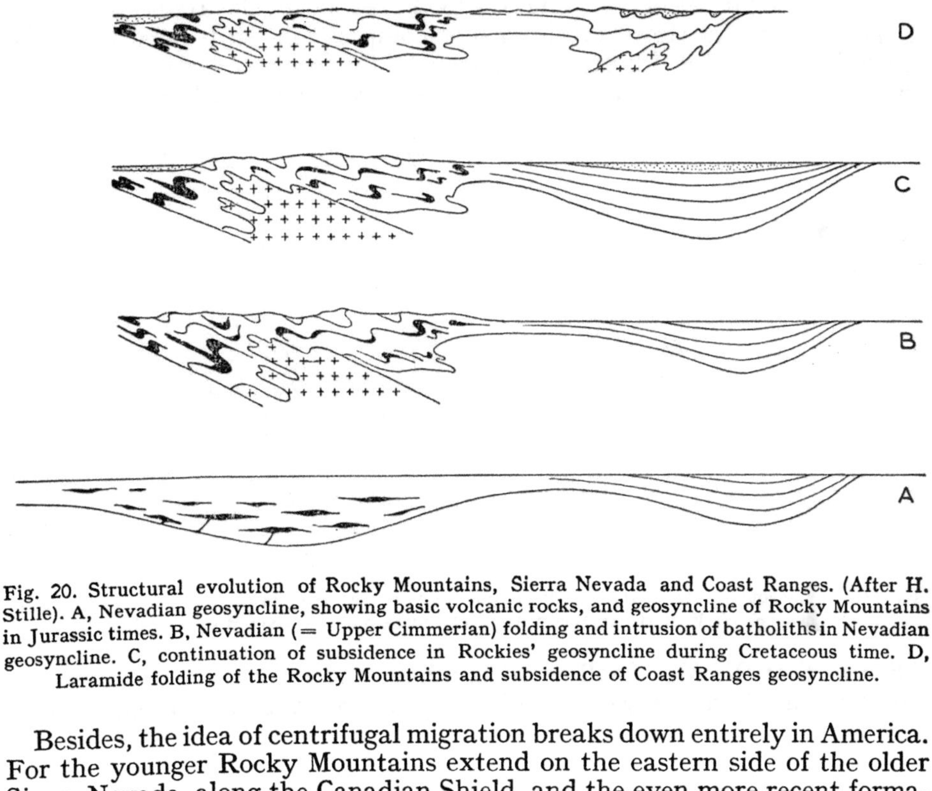

Fig. 20. Structural evolution of Rocky Mountains, Sierra Nevada and Coast Ranges. (After H. Stille). A, Nevadian geosyncline, showing basic volcanic rocks, and geosyncline of Rocky Mountains in Jurassic times. B, Nevadian (= Upper Cimmerian) folding and intrusion of batholiths in Nevadian geosyncline. C, continuation of subsidence in Rockies' geosyncline during Cretaceous time. D, Laramide folding of the Rocky Mountains and subsidence of Coast Ranges geosyncline.

Besides, the idea of centrifugal migration breaks down entirely in America. For the younger Rocky Mountains extend on the eastern side of the older Sierra Nevada, along the Canadian Shield, and the even more recent formations of the Coast-Ranges emerge on the other side of the Sierra Nevada. Such basins as Pugget Trough and California Valley formed again on the Canadian side of the Coast-Ranges. These examples make it quite clear that the centrifugal displacement of zones of folding confines itself to definite areas. Even in those regions where a centrifugal migration of zones occurs in the major systems, the following phenomenon can be observed (in the Appalachians e.g., and the European Variscides): a young trough of sedimentation originates on the continental side and is added as a last element to the folded chain. Moreover, intramontane troughs originate almost simultaneously. These phenomena will be discussed at greater length in the next chapter, in which attention will likewise be paid to the numerous basins and trough-shaped depressions on the continents.

Mountain-building

Not all epochs of compression observed in a mountain's structure are followed by mountain-building elevation of any importance. The eocene "flysch" of the Alps furnishes ample proof that some folded chains were raised and subjected to the influence of intensive denudation after an Upper-Cretaceous phase. The miocene "molasse" was the product of the sub-aerian denudation of ranges which had been elevated following an Oligocene epoch. However, intensive compression still persisted subsequently, during the Pliocene, and lead to the severe overthrusting of the Helvetian nappes (fig. 21). Nevertheless, it was not until the close of the

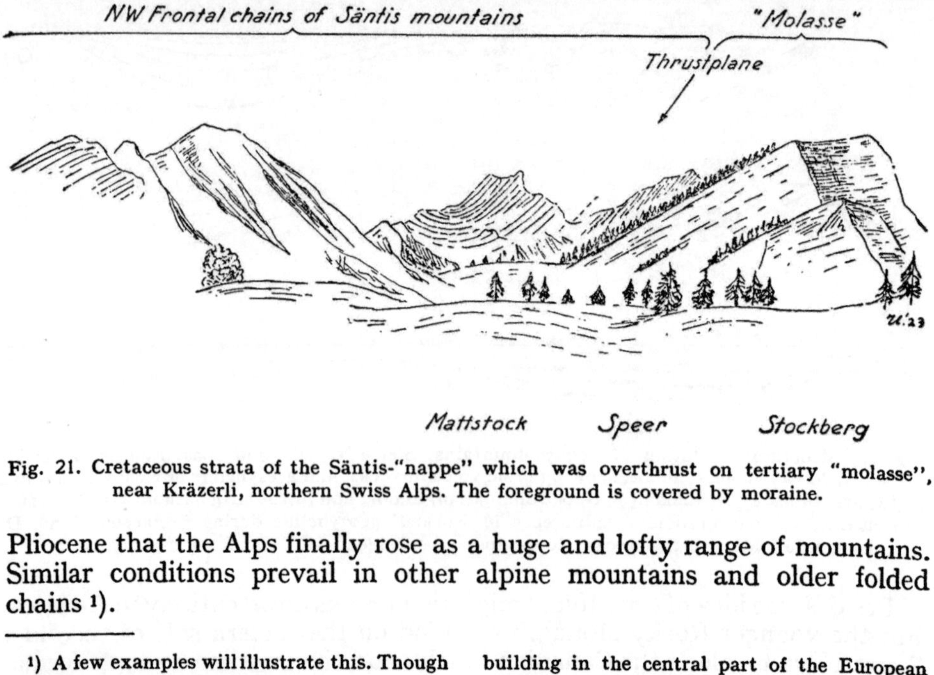

Fig. 21. Cretaceous strata of the Säntis-"nappe" which was overthrust on tertiary "molasse", near Kräzerli, northern Swiss Alps. The foreground is covered by moraine.

Pliocene that the Alps finally rose as a huge and lofty range of mountains. Similar conditions prevail in other alpine mountains and older folded chains [1]).

---

1) A few examples will illustrate this. Though several epochs are known to occur in the Caledonian belt of Spitsbergen, Scandinavia and Scotland, mountain-building — i.e. a considerable rise above sea-level accompanied by erosion — is known to have taken place towards the close of the Silurian, succeeding the Ardennian phase in Spitsbergen and probably the Ardennian or Erian in England and Scandinavia. The same occurred in the North-American Caledonides after the taconic revolution. The most important period of mountain-building in the central part of the European Variscides was posterior to the Sudetic epoch. In the northern, coal-bearing sector of Europe, however, and in Asturia, the most important mountain-building followed the Asturian phase, and an even later epoch — namely the Saalian, — in the Pyrenees and the Ural Mountains. The period of variscian mountain-building varies in different parts of America. The Arbuckle-Wichita geosyncline was elevated after the Arbuckle period of compression and had already been peneplained to a considerable

Besides, many examples show that the process of elevation may subsequently be repeated in a folded and peneplained system. This second elevation, or "rejuvenating" movement, of the already considerably peneplained older chains, which continues from the late Tertiary up to our own day, has brought about the present aspect of the Rocky Mountains, including that of the Sierra Nevada of North-America and the South-American Andes and must be held responsible for the disparity between the deeply eroded structures of the latter, with their many intrusions of batholiths, and those of other mountains, in which the last principal epoch of folding and subsequent elevation was of a more recent date (cf. the Alps) [1]).

All available data show that the largest expanse of mountain-chains since Cambrian time was occupied by the Variscides (Pl. 2), especially towards the Upper-Carboniferous (post Sudetic and pre Mid-Permian). The second largest area has been occupied by the Alpine-system since the Pleistocene (Pl. 4). The Caledonides (Taconic and Silurian), extending over large tracts, especially in the continents of the northern hemisphere (Pl. 1), were less extensive than the Alpine belts. The Mesozoic Mountains covered the smallest areas, and surround particularly the Pacific (Pl. 3). The Variscides and Alpine chains also show a circum-Pacific arrangement, but these belts likewise occupied large areas in an E. W. direction, along the southern side of the northern continents. Mountains have a very strong influence on atmospheric circulation, and this influence must therefore have been exceptionally severe in the Upper-Carboniferous and Pleistocene (this is illustrated by the schematic graph in Table II).

Small arrows in Plates 1, 2 and 4 indicate the possible trend of the Caledonides, Variscides, and Alpine systems, whose chains are intersected by the coasts of the Atlantic. This question will be discussed extensively in the Appendix and will moreover be dealt with in Chapter VI, which will include an examination of the problems of the oceanic sectors.

---

extent before being overrun by the Ouachitas during the Permian (Appalachian epoch). A large part of the Appalachians had already become an area of erosion after the Bretonic epoch, though the marginal deep was folded and overthrusted only during the Appalachian phase. In South-America the final folding and subsequent elevation occurred towards the close of the Paleozoic. In Asia, variscian folding around the Angara Shield took place in extensive areas at a relatively early date (Bretonic and Sudetic), and large mountains must consequently already have existed as early as the Upper-Carboniferous. On the other hand, the variscian zones in Taimyr, the Ural and the southern and south-eastern sector of Asia appear to have originated as late as the Permian. The Australian Variscides, too, only originated as a large unit during the Permian, though one zone in the north-east may have formed at an earlier date. The final epoch of the formation of the Australian Caledonides dates from the Upper-Devonian. The nevadian folding of the geosyncline of the Sierra Nevada was more than likely succeeded by mountain-building, so that detritus was transported from this area both eastward and westward, where it aided the sedimentation in the geosyncline of the Rocky Mountains. Opinions differ as regards the previous existence of large lower-cretaceous mountains. Crickmay was particularly doubtful in this respect. (Waters and Hedberg p. 24). It can probably also be assumed that no mountains were formed in the Rocky Mountains after the Laramide epoch. Hedberg and Waters even speak of a "low swampy plain across which the rivers from the worn down roots of the Nevadian Mountains wandered and laid down extensive flood deposits".

[1]) I cannot, therefore, endorse Gerth's view that the characteristic features of the South-American Cordilleras can only be explained by a westward drift of the continent.

Subsided blocks

The presence of detritus in some folded chains can only be explained if the previous existence is assumed of areas of denudation which have since foundered deeply below sea-level. The most important submerged areas in maps 1—4 appear east of the Appalachians ("Appalachia"), west of the Sierra Nevada ("Cascadia"), west of Spitsbergen and Scotland ("Scandia"), and, finally, south-east of Asia and east of Australia ("Melanesia"). These will now be treated accordingly.

Boesch is of opinion that the coastal area of Nova Scotia represents the only visible portion of the hinterland of the Appalachian geosyncline which is commonly known as Appalachia or Nova Scotica (Pl. 2). It is impossible to estimate the size of this hinterland, but there certainly must have existed a mountainous area large enough to supply the Appalachian geosyncline with sediments. For not only was the foreland (consisting of the Laurentian Plateau, an extension of the Canadian Shield) covered by the sea, but Barrell was able to reconstruct an enormous devonian delta in the Appalachians, and its structure shows that all the material came from the east (fig. 22). The very valuable seismic observations of Ewing and his collaborators reveal that the eastern part of Appalachia now lies buried from 3,000 to 4,000 meters below the shelf (fig. 23). This area would seem to have foundered more as a result of warping than of faulting. An important part of the shelf-sediments probably consist of triassic and jurassic strata. The region of the Appalachian Mountains can therefore probably be said to have remained above sea-level during these periods, and to have supplied the eastern area, which was then subsiding, with sediments. These deposits, together with cretaceous and tertiary strata, built up a considerable part of the shelf. Present data make it as yet impossible to determine whether the shelf's steeper slope should be regarded as a fault or as the outer margin of a delta. If "Appalachia" can be assumed to have extended some 200 miles beyond the present coast (a conclusion arrived at both by Barrell and Schuchert), thick paleozoic sediments — the erosion products of Appalachia — ought at any rate to lie buried in the Atlantic east of the shelf.

Some geologists believe that a land-area south of the Ouachitas supplied the homonymous geosyncline with detritus in the same way as Appalachia had formerly supplied the Appalachian geosyncline with erosion products. This borderland was called Llanoria (Pl. 2). Van Waterschoot van der Gracht regards the Ouachitas as the mere frontal ranges of a far more important and larger system, most of which would at present lie beneath the Gulf Coast basin.

The geosyncline and folded chains of the Sierra Nevada should not be regarded as extending along the edge of the Canadian Shield, as these formations verge upon a borderland known as Cascadia (Pl. 3). The existence of a former area in the west was also deduced by Brock from the facies of paleozoic and mesozoic sediments in Canada. That such a mountainous area had existed in the west during the Mesozoic was a logical consequence, according to Born, of observations in this area, for the mesozoic sediments of the geosyncline from which the Coast Ranges and the Sierra Nevada subsequently emerged, could not possibly have come from the east, since a wide zone, extending as far as — and including — the Rocky Mountains lay beneath the sea at that time. It seems highly improbable that the geosyncline of the Sierra Nevada would have been supplied with sediments from an area

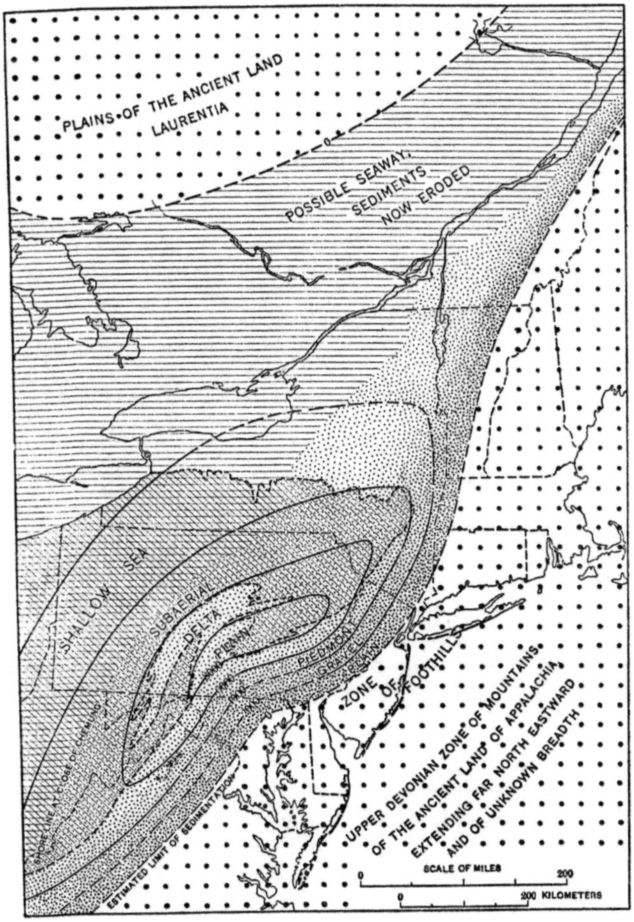

Fig. 22. Reconstruction of a devonian delta of the Appalachian geosyncline (after J. Barrell). The contours show original thickness of sediments. The diagonal lines indicate the area where the deposits still occur. The small dots represent fresh-water and brackish-water deposits; the horizontal lines marine strata; and the large dots areas of denudation.

as far distant as the Canadian Shield, especially since the sediments had to be transported over a submarine ridge between the Nevadian and the Rockies' geosynclines (see fig. 20).

Holtedahl pointed out that the devonian and tertiary sediments of

Spitsbergen originated in an area further west, known as Scandia (Pl. 1). The latter must have vanished into the Atlantic at a time when the tertiary strata of Spitsbergen were being folded. The mesozoic strata provide another indication of the presence of an area of denudation in the west, according to Frebold, and both authors agree that this land cannot be identified as East Greenland, nor that it could have been connected with the same [1]). The only remaining part of Scandia, or the North Atlantic Continent, as it is sometimes called, may perhaps be the precambrian area of the Hebrides and North-West Scotland, against which the Scottish Caledonides were overthrust (see fig. 14, 17 and Pl. 5). But the existence of a precambrian block west of Spitsbergen and extending as far as Scotland need not imply that it occupied the whole North Atlantic [2]).

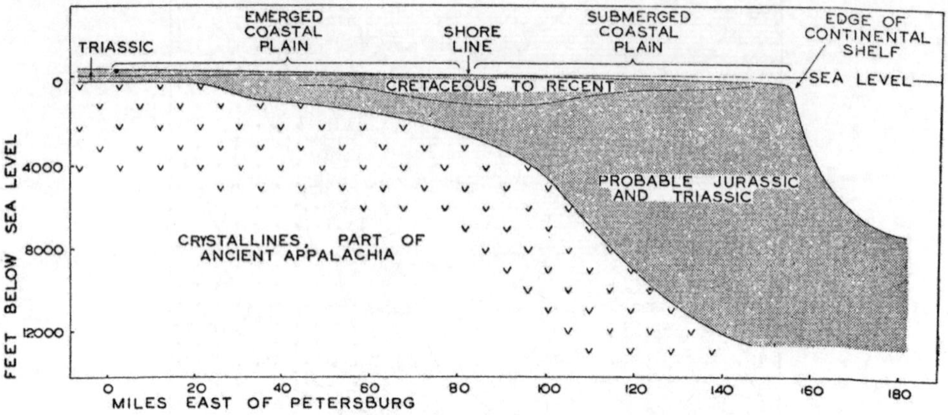

Fig. 23. Section across the Atlantic coastal plain of North-America from Petersburg op to 180 miles East of this locality. Based on seismic refraction measurements of Ewing, Crary and Rutherford (After B. L. Miller).

That a former mountainous area existed along the West Coast of Africa can be deduced from the distribution of the sediments in the Congo basin, which grow steadily coarser towards the west [3]).

The Melanesian Islands represent mere fragments of an extensive land-area. The presence of mesozoic sediments in New Caledonia can only mean that important areas of erosion, all of which have since disappeared, were present in the immediate neighborhood. Besides, it might be asked whether it would be possible to indicate at present some area of denudation which would formerly have supplied the tertiary sediments of Viti Levu with erosion products. The same applies to all those cases in which large quantities of pretertiary and tertiary detritus are found on small islands. We find it difficult to picture the formation of these rocks under the prevailing distribution of land and sea. These facts, including other data, (such as petrographical) have lead to the wide-spread assumption that the above

1) Frebold, 1935, p. 177.
2) see e.g. Holtedahl, 1920, p. 18, fig. 12.
3) Veatch, 1935, p. 57, 156.

area should be regarded as a continent, or, in other words, as a sialic block, large parts of which have subsided to a considerable depth. Geologists often refer to it as the Melanesian Continent (Pl. 4).

Fig. 24 shows Melanesia's eastern and southern boundaries. The submarine relief of this block is extremely irregular at present. Many islands and submarine ridges show a linear or arcuate arrangement. Several authors believe that the preceding characteristics, including the long deep-sea troughs, prove that important crustal folding has also affected the submerged part of Melanesia. The deepest deep-sea trough (with a depth of more than

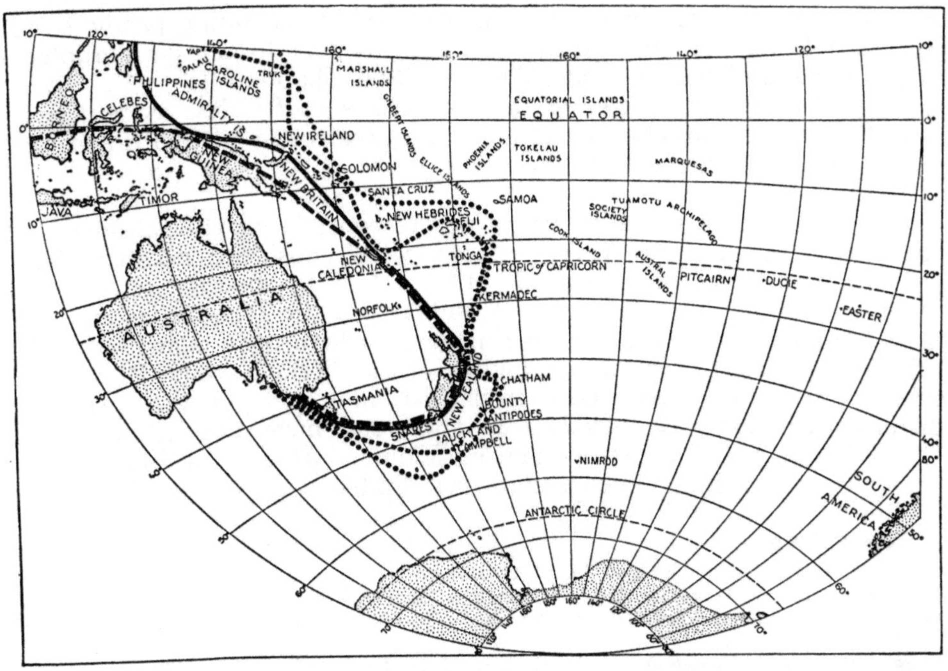

Fig. 24. Melanesia and the Pacific basin (From H. Ladd): .... probable boundary of former Melanesian continent; .... outlines area showing metamorphic or plutonic rocks; .... outlines area within which paleozoic rocks are present; .... outlines area showing mesozoic rocks.

10,000 meters east of the Philippines) is found in this area, though even in this case "with islands containing rocks of definitely continental types such as granite, syenite, serpentine and schists on both sides of it". Faulting probably constituted another important feature of the foundering, or submersion, of large parts of the Melanesian Continent. Many geologists in distant parts of this region arrived independently at a similar conclusion [1]). Chubb [2]) describes the process as "an area of continental land, that was gradually folded, fractured and submerged in Mesozoic and Tertiary times".

The exact period of the "foundering" has not been ascertained in all

1) Ladd, 1934, p. 51.

2) Chubb, 1934, p. 289. See also Schuchert 1926, p. 101.

cases, but there are indications that in one area — in the south-eastern marginal zone — important faulting occurred during the Upper-Tertiary, while in Viti Levu, to mention but one example, major movements were observed along the faults towards the close of the Tertiary, and these are thought to have accompanied the foundering of large parts of Melanesia [1]).

Melanesia's south-eastern boundary consists of a linear arrangement of volcanoes, and it is highly probable that the accumulation of these magmatic products is closely associated with a deep fault plane. This belt can be followed from New-Zealand to the neighborhood of Samoa, and constitutes an important line of demarcation of volcanic rocks, known as the andesite-line. The subsided continent, as compared to the remaining pacific sector east of Melanesia, is now known as an andesite zone from a petrographical point of view and will be discussed more fully in Chapter VI.

We will confine ourselves to these examples. A few less evident cases (e.g. Choco and Burckhardtland along the west coast of South-America) will be discussed in the Appendix.

It is difficult to say what influence was responsible for the subsidence of the blocks, but in a few areas the periods in which subsidence began, indicate that it can probably be associated with events in neighboring geosynclinal zones. Appalachia began in fact to subside during the Triassic, i.e. after the mountain-building of the neighboring Appalachian geosyncline. Cascadia subsided after the upper jurassic folding of the Sierra Nevada. Melanesia's submersion seems to be related to the laramide and younger periods of disturbance in South-Eastern Asia. Scandia would have subsided during the Upper-Tertiary, and contemporaneous folding (including faulting accompanied by volcanism) is known to occur in Spitsbergen.

References

Publications of general importance.

Born, A. *Der geologische Aufbau der Erde* (Handb. d. Geophysik Bd. II, 1932).

Born, A. *Über Werden und Zerfall von Kontinentalschollen* (Fortschritte der Geologie und Palaeontologie, Bd. X, 1933).

Bucher, W. H. *The deformation of the Earth's Crust* (Princeton University Press, 1933).

Kossmat, Fr. *Paläogeographie und Tektonik* (1936).

Schaffer, F. X. *Lehrbuch der Geologie* (III, Geologische Länderkunde, Regionale Geologie 1941).

Stille, H. *Grundfragen der Vergleichenden Tektonik* (1924).

Stille, H. *Bemerkungen betreffend die „sardische" Faltung* (Zeitschr. Deutsch. Geol. Gesellsch. 91, 1939).

Suess, Ed. *The Face of the Earth* (*Das Antlitz der Erde*), transl. Sollas, vol. I—IV, 1904—1909).

Europe, Spitsbergen and Greenland.

Bailey, E. B. and Holtedahl, O. *Northwest Europe, Caledonides* (Reg. Geol. d. Erde, Bd. 2. II, 1938)

Bubnoff, S. von, *Geologie von Europa* (Geol. d. Erde 1, 2, 1926—1936).

Cadisch, J. *Das Werden der Alpen im Spiegel der Vorlandssedimentation* (Geol. Rundschau 19, 1928).

Cornelius, H. P. *Zur Vorgeschichte der Alpenfaltung* (Geol. Rundschau 16, 1925).

Groenland 1939. *Tagung der Naturf. Gesellsch. Schaffhausen* (XVI, 1940).

Frebold, H. *Geologie von Spitsbergen, Bäreninsel, etc.* (Geol. d. Erde, 1935).

Holtedahl, O. *Paleogeography and diastrophism in the Atlantic Arctic Region during paleozoic time* (Americ. Journ. Sci., 49, 1920).

---

[1]) Ladd, 1934, p. 51.

JONES, O. T. *On the evolution of a geosyncline* (Quart. Journ. Geol. Soc. of London 94, 1938).
SEYDLITZ, W. v. *Diskordanz und Orogenese der Gebirge am Mittelmeer* (1931).
SCHWINNER, R. *Mikroseismische Bodenunruhe und Gebirgsbau* (Zeitschr. f. Geophysik, 9, 1933).
TEICHERT, C. *Geology of Greenland* (Geology of North America, Geol. d. Erde I, 1939).
WATERSCHOOT VAN DER GRACHT, W. A. J. M. VAN, *A structural outline of the Variscan front and its foreland from South Central England to Eastern Westphalia and Hessen* (C. R. 2d Congr. Stratigr. Carb. Heerlen 1938).
WATERSCHOOT VAN DER GRACHT, W. A. J. M. VAN, *The Paleozoic geography and environment in North-western Europe as compared to North-America* (Ibidem p. 1387—1429).

North- and Central-America.
BARRELL, J. *Upper Devonian delta of the Appalachian geosyncline*, (Americ. Journal of Sci. 27, 1914).
BOESCH, H. H. *Zur Geologie des östlichen Nordamerika* (Eclog. Geol. Helvet. vol. 32, 1939).
BROCK, R. W. *Structure of the pacific region of Canada* (Transact. P. Pacific Sci. Congr. Australia 1923, vol. 1).
EWING, M.; CRARY, A. P., RUTHERFORD, H. N. and MILLER, B. L. *Geophysical investigations in the emerged and submerged atlantic coastal plain* (Bull. Geol. Soc. America, vol. 48, 1937).
GERTH, H. *Die Fortsetzung der Venezuelanische Kordilleren in den Antillenbogen* (Geol. Rundschau 31, 1940).
KING, P. B. *An outline of the Structural Geology of the United States* (Guide Book 28, XVI Int. Geol. Congres 1933).
REED, R. D. and HOLLISTER, J. S. *Structural History of California* (Bull. Americ. Assoc. of Petrol. Geol. 1936).
RUEDEMANN, R. and BALK, R. *Geology of North America* (Geol. d. Erde, vol. I, 1939).
RUTTEN, L. *Oude Land- en Zeeverbindingen in Midden Amerika en West Indië* (Tijdschr. Kon. Ned. Aardrijksk. Genootsch. 1934).
RUTTEN, L. *Über den Antillenbogen* (Proc. Kon. Acad. v. Wet. Amsterdam 38, 1935).
SAPPER, K. *Mittelamerika* (Handb. Reg. Geol. 8, 1937).
SENN, A. *Paleogene of Barbados and its bearing on history and structure of the Antillean Caribbean region* (Bull. Americ. Assoc. of Petrol. Geol. 1940).
SCHUCHERT CH. *Historical Geology of Antillean-Caribbean Region* (1935).
STILLE, H. *Einführung in den Bau Amerikas* (1940).
WATERS, A. C., and HEDBERG, H. D. *The North American Cordillera and the Caribbean Region* (Reg. Geol. d. Erde 3, IVa, 1939).
WATERSCHOOT VAN DER GRACHT, W. A. J. M. VAN, *The Permo-Carboniferous Orogeny in the South Central United States* (Verh. Kon. Acad. v. Wet. Amsterdam 27, 1931).
WEYL, R. *Zum Bau des Antillenbogens* (Zentrallbl. Miner. etc. Abt. B, 1940).

South-America and Antarctica.
BAKER, H. A. *Final Report on geological investigations in the Falkland Islands* (1923).
GERTH, H. *Geologie Sudamerikas* (Geol. d. Erde I, 1932, II, 1935, III, 1941).
GERTH, H. *Die Kordilleren von Südamerika* (Reg. Geol. d. Erde, Bd. 3, IVb, 1939).
NORDENSKJÖLD, O. *Antarctica* (Handb. Reg. Geol. 1913).
STEINMANN, G. *Zum Bau der östlichen Pazifik* (Geol. Rundschau 20, 1929).
STILLE, H. *Die Entwicklung des Amerikanischen Kordilleren-systems* (Sitzungsber. Preuss. Akad. Wiss. XV, 1936).
STILLE, H. *Einführung in den Bau Amerikas* (1940).
TAYLOR, G. *Antarctica* (Reg. Geol. d. Erde 1, VIII, 1940).
WILCKENS, O. *Der Südantillenbogen* (Geol. Rundschau 24, 1933).

Africa and Arabia.
Carte géologique internationale de la Terre. (publ. Preuss. geol. Landesanstalt, Feuille 62, 63, 68, 69, Afrique).
DU TOIT, A. *Geology of South Africa* (Sec. ed. 1939).
HENNIG, E. *Afrika* (Reg. Geol. d. Erde I, V, 1938).
KRENKEL, E. *Geologie Afrikas* (Geol. d. Erde, 1—3, 1925—1938).
LAMARE, P. *Structure geologique de l'Arabie* (Ed. Béranger, Paris, 1936).

Asia.
ARGAND, E. *La tectonique de l'Asie* (C. R. XIIIe Congr. Geol. Intern. 1922).
ARNI, P. *Relations entre la structure régionale et les gisements minéraux et petrolifères d'Anatolie* (Bull. de l'Inst. et de Recherches Minières de Turquie 2, 1939).

Baier, E. *Ein Beitrag zum Thema Zwischengebirge* (Zentralbl. f. Miner. Abt. B. 1938).
Baier, E. *Das Iranische Binnenland* (N. Jahrb. f. Min. Abt. B. 1940).
Blanckenhorn, M. *Syriën, Arabiën und Mesopotamien* (Handb. Reg. Geol. 1914).
Brouwer, H. A. *The Geology of the Netherlands East Indies* (1925).
Brouwer, H. A. *Geological expedition to the Lesser Sunda Islands* (I, II, 1940; III, 1941).
Chhibber, H. L. *The Geology of Burma* (1934).
Cotter, G. de P. *The Indian Peninsula and Ceylon* (Reg. Geol. d. Erde I, VI, 1938).
Furon, R. *Geologie du Plateau Iranien* (Mém. Musium Nat. Hist. Nat. 1941).
Geological atlas of Eastern Asia (Tokyo Geogr. Soc. 1929).
Geological atlas of the Malay archipelago (Tokyo Geogr. Soc. 1932).
Geological map of the Union of Soviet Soc. Rep. 1 : 5.000.000 (1937).
Gregory, J. W. *The structure of Asia* (1929).
Hanzawa, S. *Topography and Geology of the Riukiu Islands* (Sci. Rep. Tohoku Imp. Univ. 17, 1933).
Hanzawa, S. *Geolog. History of the Riukiu Islands* (Proc. Imp. University 11, 1935).
Hayasaka, I. and Takahashi, H. *An outline of the Geology and mineral resources of Taiwan* (1929).
Heim, A. and Gansser, A. *Central Himalaya* (Geolog. Observ. Denkschr. d. Schweiz. Naturf. Ges. 73, 1, 1939).
Heim, A. and Hirschi, H. A. *Section of the Mountain Ranges of North-Western Siam* (Eclog. Geol. Helvetiae 32, 1939).
Kanehara, N. *The Geology and mineral resources of the Japanese Empire* (1926).
Kobayashi, A. *A sketch of Korean Geology* (Americ. Journ. Sci. 26, 1933).
Lee, J. S. *The Geology of China* (1939).
Leuchs, K. *Geologie von Asiën* (I, 1, 1935, I, 2, 1937).
Leuchs, K. *Geologische Entwicklung von Anatoliën* (Leipziger Vierteljahrschr. für Südosteuropa 2—2, 1938).
Obrutschew, W. A. *Geologie von Sibirien* (Fortschr. d. Geol. und Pal. 15, 1926).
Oswald, F. *Armenien* (Handb. Reg. Geol. 1912).
Ozawa, Y. *The post-paleozoic and late mesozoic earth movements in the inner zone of Japan* (Koto Commemoration volume 1925).
Pascoe, E. H. *Geological map of India and adjacent countries* (5ă ed. 1931).
Philippson, K. *Kleinasien* (Hand. Reg. Geol. 1918).
Rutten, L. M. R. *Voordrachten over de Geologie van Nederlandsch Oost Indië* (1937).
Scrivenor, J. B. *The Geology of Malaya* (1931).
Stahl, A. F. von, *Persien* (Handb. Reg. Geol. 1915).
Stahl, A. F. von, *Kaukasus* (Handb. Reg. Geol. 1923).
Umbgrove, J. H. F. *Geological History of the East Indies* (Bull. Americ. Assoc. Petrol. Geol. 22, 1938).
Yabe, H. *Larger geotectonics of the island arc of Japan* (Proc. Imp. Acad. Japan 5, 1929).

Australia, New Zealand and Melanesia.
Andrews, E. C. *The structural History of Australia during the Paleozoic* (Journ. and Proc. of the Royal Soc. of New S. Wales 71, 1938).
Benson, W. N. *The structural features of the margin of Australasia* (Transact. N. Zealand Inst. 55, 1924).
Chubb, L. J. *The structure of the Pacific Basin* (Geol. Magaz. 71, 1934).
David, T. W. E. *Explanatory notes to accompany a new geological map of Australia* (1932).
Hoffmeister, J. E. *Geology of Eua-Tonga* (Bernice P. Bishop Museum, Bull. 96, 1932).
Ladd, H. *Geology of Viti Levu* (Bernice P. Bishop Museum. Bull. 119, 1934).
Marshall, P. *Oceania* (Handb. Reg. Geol. VII, 2, 1911).
Marshall, P. *Stability of lands in the S. W. Pacific* (Rep. of the Australian and New Zealand Assoc. for the Advanc. of Sci. 21, 1933).
Mawson, D. *The Geology of the New Hebrides* (Proc. Linn. Soc. Nw. S. Wales 30, 1905).
Schuchert, Ch. *The problem of continental fracturing and diastrophism in Oceanica* (Americ. Journ. Sci. 47, 1916).
Wilckens, O. *Stratigraphie und Bau von Neu Kaledonien* (Geol. Rundschau 16, 1925).

CHAPTER III

BASINS AND TROUGHS

"The breaking up of the terrestrial globe, this it is we witness".
(ED. SUESS)

Introduction

Folded mountain-chains — those majestic features of the earth — have always attracted both the tourist and the geologist. There is another feature, however, which, though less stimulating and even monotonous, can be said to be of primary importance in the earth's aspect. We mean such formations as troughs and basin-shaped depressions, containing more or less folded strata or even sediments which have remained almost undisturbed. As soon as we begin to examine these structures more closely, it transpires that they present heterogeneous characteristics, revealing different shapes, sizes, structures, depths, tectonic frames, periods of formation, and a varying distribution of facies and duration of development. In studying the earth's structural history, special attention will have to be paid — as in the case of the folded chains discussed in Chapter II — to the location and time of origin of the basins, without losing sight, however, of those other diverse characteristics which have just been mentioned.

The shape of these formations ranges from long troughs and elliptical depressions to practically circular basins. The thickness of the accumulated sediments varies between a few hundred meters and several thousands in often uniform shallow facies. Some basins are consequently not unlike geosynclines both as regards their form and the distribution of the sediments' facies. It is in fact difficult to draw a sharp line between basins, troughs and geosynclines, or to formulate some definition which would distinguish clearly between them as separate physiographic units. The extremes are different enough; but the intermediate types cause difficulties. Thus we find one author calling a certain type of formation a geosyncline, while another refers to it as a basin. Basins and troughs are often bounded by faults and overthrusts, and this circumstance makes it difficult to distinguish even between a trough and a graben. To make things still more complicated, names such as "stable and unstable shelf" have since been introduced!

We do not propose to dwell upon all these different names, but will describe the features of each type of basin as it comes up for discussion and will then choose the name which seems to fit it best.

A study of continental basins and troughs has induced me to divide them into two groups, each group being subdivided as follows into two types (fig. 25):

Group 1.
- Type I marginal deep.
- Type II intramontane trough.

Group 2.
- Type III nuclear basin.
  - (a) with an isochronous frame.
  - (b) with an anisochronous frame.
- Type IV discordant basin.

It is impossible to distinguish accurately between both types in each group. To the following review of continental types will be added a discussion

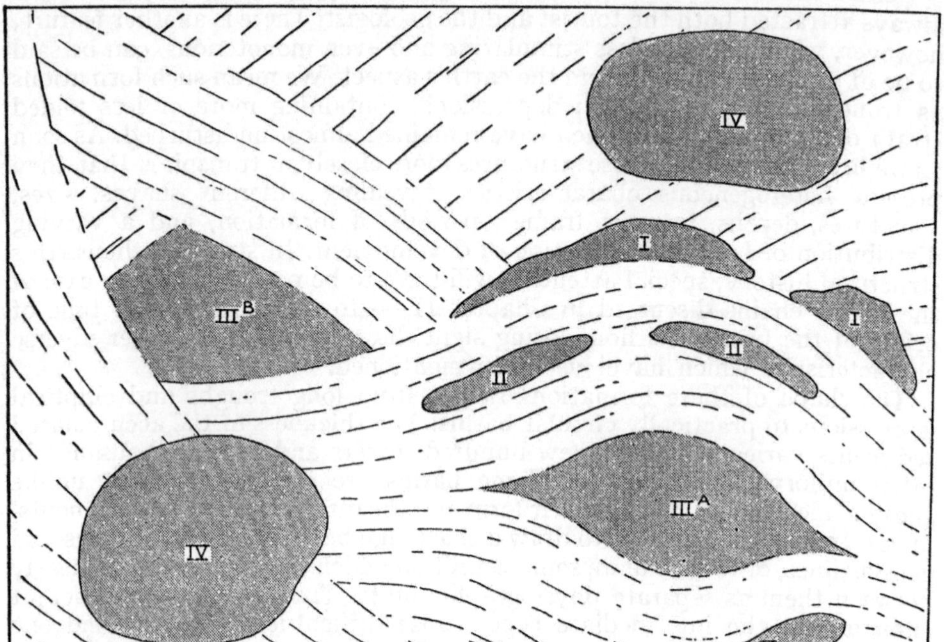

Fig. 25. Schematic review of four types of basins. The trends of folded structures are indicated by lines. I marginal deep; II intramontane trough; III nuclear basin, (A) with isochronous frame, (B) with anisochronous frame; IV discordant basin.

and a classification of deep-sea basins and troughs. At the end of the chapter attention will be drawn to the chronological relations to other phenomena in the earth's structural history.

Continental basins and troughs

The extensive Variscian mountains of Western Europe emerged towards the close of the Lower-Carboniferous (following the Sudetic epoch of compression). A relatively small area of sedimentation later formed along the chain's northern margin. The latter can be traced from Ireland to Westphalia over England, and through Belgium and Limburg. This represents the subsiding trough in which the paralic and coal-bearing Upper-Carboniferous was deposited, accumulating to a thickness of approximately 4,000 meters in Westphalia, and to as much as 7,000 meters in Upper-Schlesia. A considerable amount of detritus from the variscian mountains gathered in this trough, which constituted a marginal deep of the European Variscides. Marine Devonian and Lower-Carboniferous had already been deposited in this area previously, but the creation of the marginal deep only dates from Upper-Carboniferous times. Its contents were then folded towards the close of the Upper-Carboniferous, and during this same epoch (which was probably the Asturian), blocks of the older Variscian ranges in the south were overthrust towards the north as far as, and over, the fore-deep's sediments (fig. 26). We will not enter into details at this point, nor will we describe the important role which the Brabant Massif played during these events. We refer the reader for these questions to the references at the end of this chapter, especially to the publications of Van Waterschoot van der Gracht. The foredeep should not be imagined to be a deep-sea basin in this case. On the contrary. The enormous supply of detritus from the South kept pace with the subsidence, producing a swampy zone which was intermittently flooded by a shallow sea. It was not, therefore, a deep in the morphological sense of the word, but in the structural or tectonic sense, as Stille formulated it. The mechanism of the subsidence and sedimentation of this upper-carboniferous "geosyncline" was the object of a special study by Pruvost.

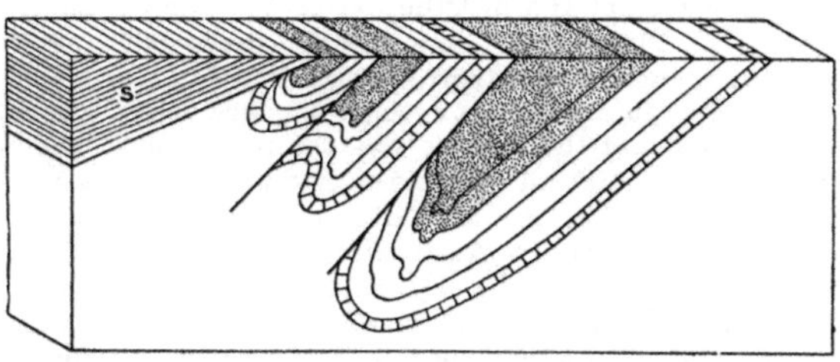

Fig. 26. Diagrammatic block of the folded and overthrusted marginal trough of the European Variscides. S Lower Paleozoic.

While discussing epochs of compression, the principal phase in the Appalachians was shown to be the Bretonic (Acadian). It involved the Eastern and most intensely folded and metamorphosed zone with its manifold intrusions of large batholiths. A marginal trough was formed along the western edge of the newly elevated mountain range following this Acadian revolution. The marginal deep, which was filled with carboniferous and permian strata, was folded by the so-called Appalachian (Saalian) phase during the Permian. This is the classical Appalachian geosyncline of former authors, for which Schuchert introduced the name "monogeosyncline". As a tectonic unit, it formed the counterpart of the marginal deep of the European Variscides mentioned above [1]).

These two examples refer to variscian marginal deeps. We will now give a brief summary of marginal deeps of the alpine belt.

A miocene trough was formed along the northern front of the Alpine chains after the intensive folding of the Oligocene. A sequence of approximately 2,500 to 3,000 meters of "molasse" was deposited into this trough, and the latter was overthrust towards the end of the Miocene (fig. 21). An identical formation — the Guadalquivir trough, which contains lower-tertiary and neogene sediments — can be observed in front of the Betic Cordillera. Its oligocene "flysch" is folded; the Burdigalian and younger strata lie undisturbed. An analogous deep is the neogene Siwalik trough, which extends along the southern margin of the Himalayas. On the Alpine chains' northern side lies the depression of Karakum. The drainage basins of the Tigris and Euphrates and the Persian Gulf form the marginal deep of the Zagros and Iran chains [2]).

Recapitulating, we may say that a marginal deep can be characterized as a subsiding trough of sedimentation, which originates along the edge of a folded chain shortly after folding and mountain-building occurred in this zone.

This trough faces the continental Shield in the European, North-American and Asiatic Variscides, and the same applies to the Guadalquivir, Molasse and Karakum troughs of the Alpine ranges. The Mesopotamian and Siwalik troughs, however, are found on the other side of the alpine belt, and an

---

1) A glance at the distribution of Variscian epochs in Plate 2 will immediately make it clear that the most southern Variscides in Asia were folded during the more recent Appalachian epoch, and that the Sudetic was the principal phase in the northern Variscides. A centrifugal displacement of epochs of folding (cf. also the same phenomenon in Australia) is what we observe here. This will be explained in the Appendix. We are as yet unable to say for certain whether we may speak in this case of a marginal deep, which would have formed on the Asiatic Variscides' southern side, and would have been folded during the Appalachian epoch, but it looks as if we might be justified in doing so.

2) Von Bubnoff also regards the area between the Carpathian Mountains and the Podolian Massif, including the zone of subsidence north of the Dobrutcha, Crimea and Caucasus, as a marginal deep (Europa 1, p. 291). This same author also looks upon the South-Russian basin as a marginal deep — in this case of the Donetz ranges, namely (ibid. p. 208).

To the above must be added the huge coal basin of Kusnezk, which continues towards the South-Schenka basin. The basin of Kusnezk lies between the caledonian Kusnezk Alatau and the variscian Salair Mountains. It began to subside during the Lower-Devonian, and 6,200 meters of neritic and paralic upper-paleozoic strata were subsequently deposited in the basin (Obrutchew even speaks of 7,500 meters of coal-bearing sediments). The basin's contents were influenced by rather intensive sudetic and upper-paleozoic folding (Leuchs 1, p. 155, fig. 48), and during this process even the cambrian strata of the caledonian Kusnezki — and the variscian Salair on its other side — were overthrust over the edge of the basin. The subsidence was resumed in the Jurassic, following the Lower-Cimmerian epoch, when 2,000 meters of Jurassic sediments were once more unconformably deposited upon the paleozoic series.

The basin's situation on the southern margin of the Asiatic Caledonides resembles that of a caledonian marginal deep. Its subsidence began in the Devonian. One remarkable thing, however, is that it is now bounded on the other side by the variscian Salair Mountains, which were folded during the Sudetic epoch (this basin represented the Salair Mountains' marginal deep subsequently to this Sudetic epoch; it was folded towards the close of the Paleozoic). Another point which should be noticed is that this subsidence was repeated a third time in the Jurassic, resulting in the formation of a trough of the intramontane type.

identical situation can be observed in the Asiatic Caledonides. We propose to give these formations a neutral name, for they can be called "foredeep" in one case, and "hinterdeep" in another, while at times both terms may apply — it all depends on the way an investigator looks at it. We shall therefore speak of them as "marginal deeps". Stille, too, refers to them as "saumtiefe". As stated above, the Appalachians form an example of what Schuchert called a "monogeosyncline".

Geosynclinal migration was discussed in the preceding chapter, and we

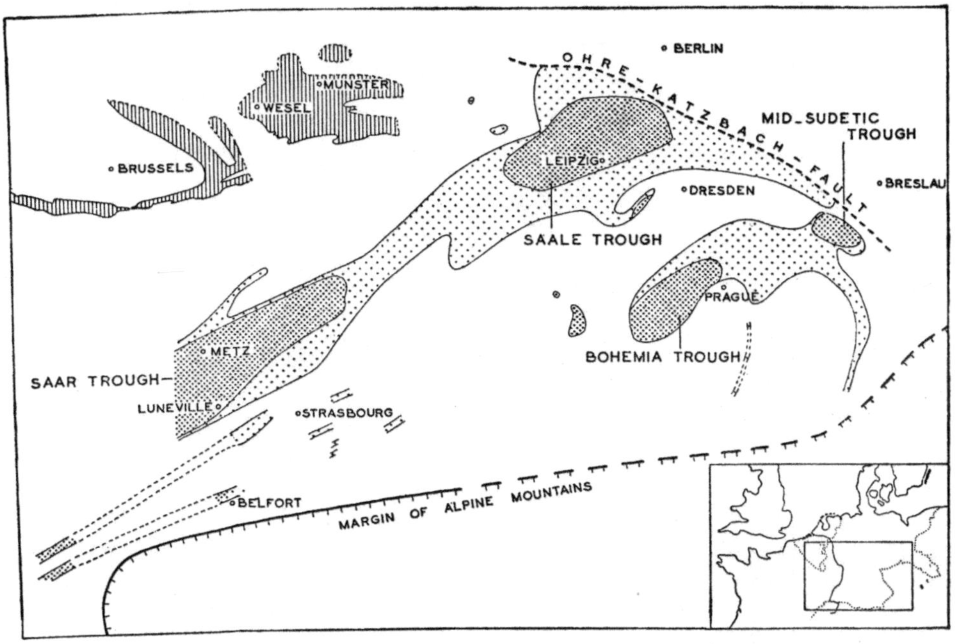

Fig. 27. Intramontane troughs in the upper-paleozoic Variscides of Europe. Where subsidence started in upper-carboniferous time areas are closely dotted, where it began in permian times areas are widely dotted; marginal deep of Variscides is indicated by vertical shading.

might be induced to speak of a repeated migration of geosynclines. We want to restrict the application of the term "marginal deep", however, to the youngest and final marginal trough in the history of a given mountain-belt.

One notable feature is that this marginal deep does in fact originate along the margin of the older folded chain, whereas the migrating geosynclines of the belt's earlier history are formed in most cases partly in the area of the preceding zone of folding and sedimentation.

Another notable feature is that the marginal deep is in many cases situated on the side facing the continental Shield, thereby opposing the general tendency of centrifugal development.

If a marginal deep were to be pictured without its contents, its shape would be that of a small and typical trough with a depth of a few thousand meters. The marginal deeps of Kusnezk and Karakum may be cited as exceptions. The latter have a more limited, basin-shaped appearance. This may merely be a secundary phenomenon, due to extensive overthrusting of the marginal areas. Thus the marginal deep of the European Variscides is also entirely "obscured" by overthrusting from the south in the western part of the north of France [1]) (see fig. 26).

Troughs of sedimentation also formed within the European Variscides shortly after the sudetic folding of this zone. Special attention was paid to these formations by Born, and the problem of their origin was later studied by Stille and von Bubnoff. These troughs have long-drawn shapes, with

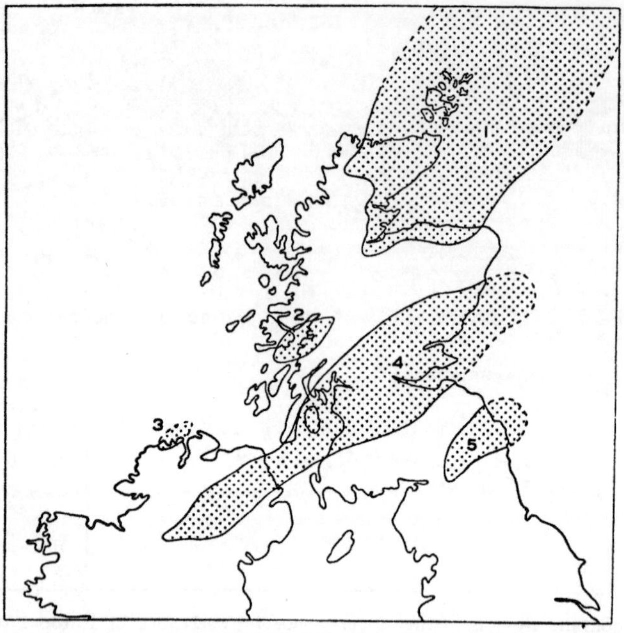

Fig. 28. Intramontane Old Red basins of Great Britain. 1. Lake Orcadie; 2. Lake Lorne; 3. Lake Fanad; 4. Lake Caledonia; 5. Lake Cheviot.

more or less elliptical and at times irregular contours. Their longest axes extend parallel with the trend of the Variscides, and ridge-shaped "geanticlinal" elevations separate them (fig. 27).

The contents of these furrows, consisting of terrigenous strata, which are in turn composed for the most part of detritus from the geanticlinal zones of the Variscides, kept up with the gradual subsidence of the bottom, and became very thick in some cases, accumulating to as much as 6,000 meters in the trough of the Saar, 2,500 meters in that of the Saale, and 5,500 meters in the Mid-Sudetic trough. Still, precise observations have shown that the pace of this subsidence was not always the same, that the troughs

1) See the tectonical map of Van Waterschoot van der Gracht, 1938.

were steadily enlarged in the direction of their longest axis, that they also broadened, and that there was a distinct relation between volcanic phenomena a hiatus in the sedimentation, and warping in the strata, especially between the middle and upper part of the Lower-Permian. The beginning of subsidence varies between the lower part of the Upper-Carboniferous and the upper part of the Lower-Permian. Another striking feature is that many troughs (e.g. the Saar trough) originated on the boundary of a crystalline and sedimentary area. Thus in several cases these furrows may in a certain sense be said to represent marginal deeps, and it is in fact impossible to distinguish clearly between Type I (marginal deep) and Type II (intramontane trough).

The intramontane troughs of the Caledonides of Scotland and Ireland,

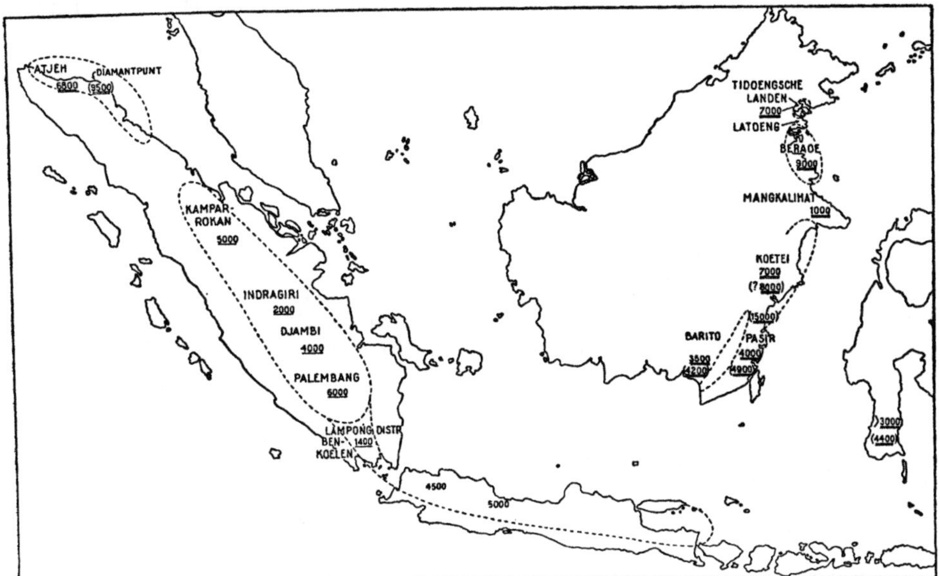

Fig. 29. Idiogeosynclines in the western part of the East-Indies. The figures between brackets indicate the thickness of the whole Tertiary, in meters. The geosynclinal subsidence began during the Miocene wherever the figures are underlined and during the Eocene where double underlined.

which were also formed in the direction of the mountains' trend, were filled with "Old Red". These troughs are known to us as "Lake", Orcadia, Caledonia Cheviot, Fanad and Lorne (fig. 28) [1]).

There are many examples of intramontane troughs in alpine zones of folding. For instance, the Maracaibo, Orinoco and Magdalena troughs in

1) The so-called basin of Gorlowsk situated West of the Salair chains furnishes an example of an intramontane trough in the Asiatic Variscides (Leuchs, 1 p. 154, fig. 45; p. 156, fig. 52). Neritic to terrestrial upper-paleozoic sediments form the contents of the trough which was rather intensely folded towards the end of the Permian, when many faults and overthrusts within the trough itself, as well as overthrusting of the framing chains in the direction of the trough came into existence, so that Leuchs even speaks of a "grabenförmige Einsenkung".

South-America, which should be regarded as such, and the Pugget trough, Ventura basin and California Valley in North-America. Other examples of intramontane troughs are the "ovas" in Asia-Minor. Mention should also be made of those basins in the East-Indian Archipelago and Burma to which I formerly referred to as idiogeosynclines, (fig. 29). In Burma, Sumatra and Java these formations are bounded on one side by a pre-tertiary area and on the other partly by a miocene zone of folding (Arakan Yoma, Barisan, South-Java) [1]).

A sharp division between marginal deeps and intramontane troughs cannot be based on the sediments' facies, nor can the intensity of movements contribute anything towards such a division. For the fact is that the facies depends on the distance between the level of sedimentation and the corresponding level of the sea. The idiogeosynclines of the East-Indies. unlike the

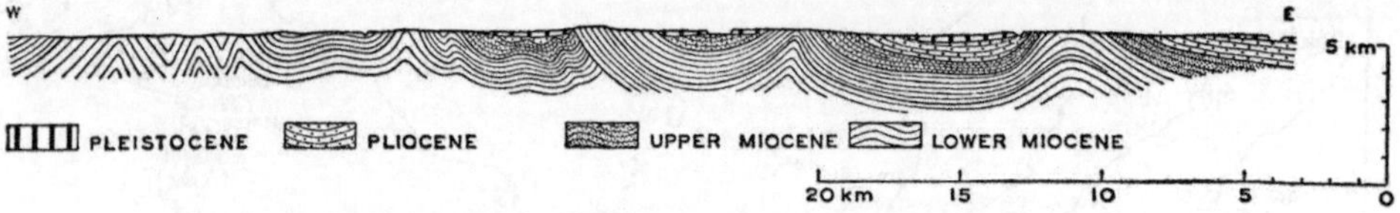

Fig. 30. Section across the geosynclinal basin of eastern Borneo. (After Jetzler).

intramontane troughs of the European Variscides, provide examples of troughs with a chiefly marine and paralic sedimentation. The frequent appearance of eruptive rocks in the intramontane troughs of Germany, and the complete, or almost complete absence of these rocks in a typical marginal deep would seem to constitute an important characteristic, but should in

---

[1]) They cannot, however be regarded as intramontane troughs of this zone, for they originated partly during the Miocene (but before the miocene folding), and partly already during the Eocene. The miocene area of compression extends to the islands west of Sumatra, and as far as the island-series formed by Timor, Tanimber, Kei and Ceram (fig. 36). The Archipelago's zone of laramide folding was probably a much wider one, and it is not improbable that it also extends over a large portion of Sumatra and Java, for cretaceous rocks are still found in the formations of the Barisan Mountains, which were folded during the pre-Tertiary, and in the exposures of pretertiary rocks in Java (1938, p. 27, 28). The idiogeosynclines of Burma, Atjeh, Djambi and Java may have a laramide basement, or they may have originated on the edge of such a basement, but whether we call them marginal deeps or intramontane troughs is just another question of names.

This is equally true of Pugget trough and California Valley, and the idio-geosynclinal troughs of Northern and Southern New-Guinea, the South of Celebes and East-Borneo. The situation becomes even more complex in this last area, for a recent deep-sea trough originated on its eastern side, as will be described later. We already mentioned the interesting Kusnezk basin when dealing with marginal deeps. This basin, which was situated on the boundary between two elements of varying ages, began by being a marginal deep of the Caledonides, then became a marginal deep of the Variscides, and finally resumed its subsiding movement as an intramontane basin after the Lower-Cimmerian phase. Mention must still be made, in this connection, of the frequent appearance of longitudinal fault zones and graben, occurring mostly in the bordering geanticlinal areas, but striking at all events in the direction of the mountains' trend. Examples are: the central graben of Spitsbergen, the rift valley in the variscian Tianchan, the „graben" in the Chilean Andes (Gerth, 1939, p. 40), the "longitudinal valley" in Sumatra and the central graben in the islands of Kei, Tanimber and Timor (1939, p. 47—49, fig. 15—17). Molengraaff referred to the pliocene graben of Timor as a geosyncline.

no case be generalized into a law, for volcanic products (viz. diabase, melaphyr and basalt) do in fact occur in the Kusnezk basin, a marginal deep of the Asiatic Caledonides. The following may be said as regards the third characteristic. The contents of the intramontane troughs of the European Variscides are hardly folded in comparison with that zone's marginal deep, but we again find examples of a rather intensive folding of intramontane troughs in the East-Indian idiogeosynclines (fig. 30).

We still have to consider the location of the troughs. The name intramontane in itself would seem to distinguish this type of trough from a marginal deep, but we saw that it is in some cases difficult to distinguish clearly between the two.

One more type of basin can be found in Europe, viz. the Hungarian, or Pannonian basin. Its basement was very probably folded during a Variscian epoch, and the basin more than likely formed a rigid nucleus around which

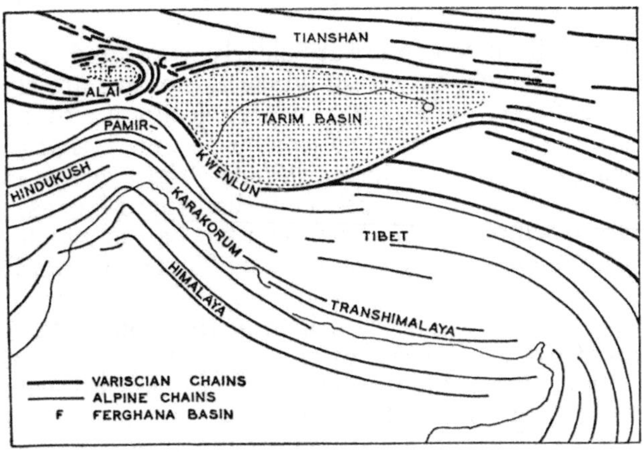

Fig. 31. Schematic illustration of the Ferghana- and Tarim basins and the surrounding mountain-chains.

the Carpathian Mountains grouped themselves in the north and east, and the Dinarides in the south and west. The Pannonian Massif itself, however, subsided, forming a basin which was filled with lower miocene up to pleistocene strata, and these were subjected to folding that adapted itself to the Alps' strike in the west, and that of the Carpathians in the east.

The precambrian nuclei of Ferghana, Tarim and Ordos played an important part during the variscian folding in Asia. The chains were pressed against, and arranged around the nuclei (fig. 31). However, the massifs later showed a tendency to subside, and became basins. Sedimentation in the basin of Tarim began during the Permian or Upper-Carboniferous and continues up to this day. The neritic, but mainly the terrestrial deposits, are probably exceptionally thick, and show several unconformities and folding of a not very pronounced character.

Type III's most salient characteristic, therefore, is that it originates on a subsiding block which, situated in a zone of folding, formerly acted as a rigid nucleus and thus produced the aforementioned effect on the trend of the folded chains that frame it. This is why I called them "nuclear basins".

The situation of the large salt basin of Texas is a different one in some respects, for it subsided on a precambrian basement, as the Llano Burnett uplift in the south-east, and the Colorado Plateau in the north-west show (see Plates 5 and 6). Its frame consists of the variscian Marathon-Ouachita chains towards the south and east, and the older Amarillo-Wichita zone in the north. The basin's former western confine is unknown, but it is now occupied by the tertiary ranges of the Rocky Mountains. This salt basin containing permian deposits up to as much as 4,000 meters) may in a certain sense be compared to Europe's Pannonian basin, but the latter's frame, unlike that of the Texas salt basin, is isochronous, as it were (the chains originated during the same period of folding). The submarine Barent-Sea Massif also has an anisochronous frame. The thickness of the deposited sediments remains unknown, but Frebold pointed out that numerous mesozoic transgressions have at any rate invaded this area.

The three types discussed so far are clearly connected with the mountain chains alongside of which, or within which the basins are situated, but the same is not immediately evident in the large group of remaining basins. We can begin by stating that the most conspicous feature is that their boundaries intersect existing structures. They originate "right across everything", so to speak, and we will therefore call them discordant basins.

Table I gives a summary of the numerous basins which belong to this group. They will be found under the heading "Type IV", and their situation is shown by Plate 6. We will discuss only one example here — viz. the basin of Paris (fig. 32). The rest will be described in the Appendix. A glance at the geological map of France shows that the basin of Paris originated in Liassic times. The deepest part gradually moved from the south-east to the north-west, lying near Paris during the Oligocene, and near the Channel coast during the Miocene. The subsidence represents approximately 1,200 meters (boring of Ferrières) in the basin's northern sector, while the precambrian basement — on which rests the Lias — descends to a depth of 1,532 meters in the "Pays de Bray". Its continuation is found in Hampshire and Wight on the other side of the Channel, and here the sediments are even considerably thicker (the boring at Portsdown reached the upper-triassic strata at a depth of 1,993 meters).

The anticlines striking west-north-west — south-south-east were formed during the Upper Cimmerian epoch. The subhercynian phase is indicated by but faint movements, the Laramide revolution by faulting and the Pyrenean epoch by "epirogenetic" movements. Finally, a rather faint folding occurred during the Attic phase. It would be wrong to think of one, large basin-shaped depression. Not only did the deepest part move west-north-west, but the bowl-shaped habitus of the basin is the result of certain transgressions and of subsequent erosion, and does not seem to reflect the morphology of the basement. Lemoine drew attention to the fact that the liassic strata of a trough extending from Lorraine in the direction of the "Pays de Bray" are thicker than those found outside this area, and remarked that this feature also manifests itself in the foundation of the mesozoic upperstructure.

An examination of the manifold discordant basins (we refer to the Appendix for details) shows that these formations originated in heterogenous basements at different periods, and that their boundaries distinctly intersect those of pre-existing structures.

Yet it can be pointed out in a few cases that a basin's original shape and boundaries are related to specific positive structural elements of their foundation. The original shape of the Liassic trough of the basin of Paris, for instance, the migration of its deepest part, and the strike of the

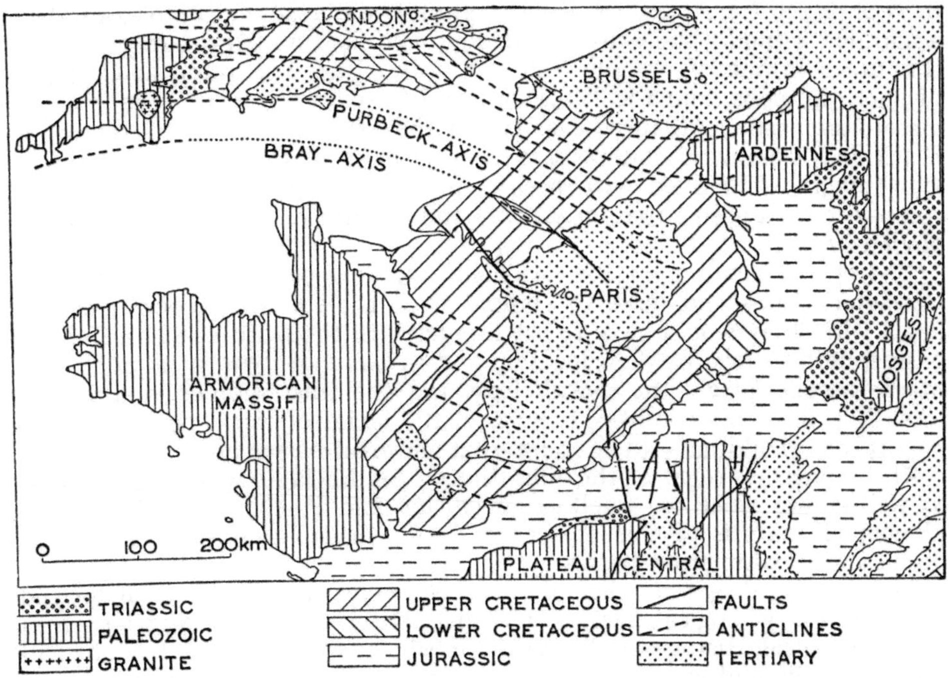

Fig. 32. Geological Sketch map of the basin of Paris and its surroundings.

anticlines, clearly reveal the posthumous activity of its variscian foundation. Similarly, the north-south boundary between the North-Sea and the West-German basin, or so-called axis of Erkelenz, and the Western boundary of the North-Sea basin — the Pennine axis, in which positive movements occurred during the Saalian epoch — probably represent very ancient trends in the structure of the basement complex. The entire arrangement of positive and negative elements in North-America is apparently also the reflection of the precambrian structure of the Canadian Shield. Africa may be cited as another example. A north-east — south-west and a north-west — south-east strike (referred to respectively as the Somali and Erythrean trend) is inherent to this continent's precambrian structure, which repeatedly manifests itself in one form or another, revealing itself also in the basins, their arrangement and boundaries.

These facts show that a discordant basin's original site, if examined more closely, will probably reveal a certain "concordance" with some of the structural elements of its surroundings, and the division of basins into nuclear and discordant basins consequently appears to be not so much a fundamental one, as one of degree.

Deep-sea basins

A few submarine basins have been mentioned among the examples given in Table I and the Appendix. These include the Persian Gulf (an example of a marginal trough), the Gulf of Martaban and Madura Straits (representing the intramontane type). The Barent-Sea and Kara-Sea stand as examples of nuclear basins, and the Black-Sea and Caspian-Sea have been cited as discordant basins. All these cases, with the exception of the Caspian and Black-Sea, deal with shallow seas, and as most of the continental basins and troughs were filled with neritic, paralic and terrestrial sediments, a comparison between these formations and the above submarine furrows is obviously warranted.

If deep-sea basins are to be classified according to our four types, their characteristics would have to comply with the following conditions. To begin with, they would have to correspond with the features of continental basins, both as regards their location and the time of their formation. And, in the second place, the morphology and the surroundings would have to satisfy the same conditions. Finally, the facies of the deposits in these depressions would have their counterpart in fossil basins on the continents.

The submarine relief of the East-Indian Archipelago, which has become so well-known since the Snellius expedition and Kuenen's excellent morphological analysis of this area, would seem a suitable subject for the discussion of submarine basins.

The present deep-sea relief of the East-Indies must have been formed in the recent geological past. I treated this question at greater length in "the Geological History of the East-Indies", as well as in a later paper of the year 1938, and will therefore confine myself here to a few of its salient features. I want to mention in the first place that the trend of the folded miocene strata is intersected at an angle by the present boundaries of the deep basins in several places from which it follows that the adjacent submarine relief must have originated at least after that miocene folding. Molengraaff explained, for example, that the Amanuban mountain-chain, in Timor, is intersected at an angle of 12° by the coast. This author's view that the origin of the deep-sea basins, and the elevation of the series of islands between them, probably occurred simultaneously, is now generally accepted.

It is difficult to determine the exact time of the beginning of these submersions. It may perhaps be placed in the Pliocene. However, the upper neogene strata of the island Buton were folded with a trend which the neighboring Gulf of Bone intersects at right angles (fig. 33). The time of the basin's formation can in this case consequently be fixed as uppermost Tertiary or Pleistocene. And if the submersion of these basins and the elevation of the intervening series of islands can be said to be closely con-

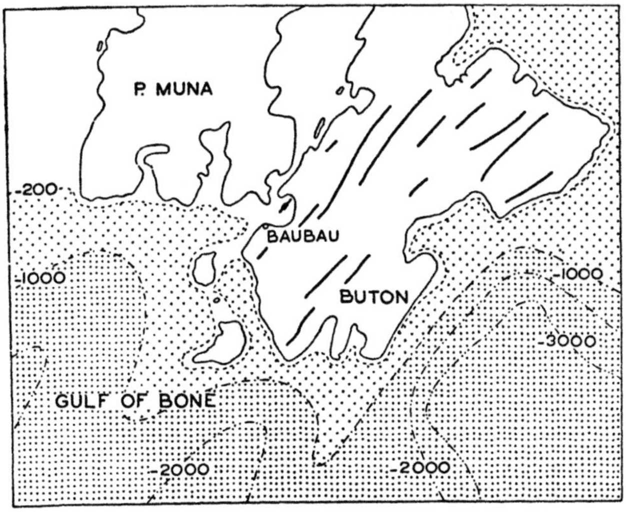

Fig. 33. Upper Tertiary anticlines in the island Buton, East-Indies.

nected, which seems probable, we are able to conclude that the most important part of the submerging movement must also have taken place in the Pleistocene in the case of the other basins. There can be no doubt that the rising movement of the islands occurred in the Pleistocene. This most recent movement, to mention but one example that could be supplemented by many others, brought parts of mountains in Central-Ceram (which lay below sea-level during the sedimentation of the marine pliocene strata in the Masiwat-Bobot graben) to a height of at least 3,000 meters above the sea. The recently elevated reef limestones and terraces give a particularly good idea of the amount and intermittent character of these recent movements. Fig. 34 gives some figures for part of the southern Moluccas. The actual pleistocene age of the elevated reef terraces has been verified in some cases.

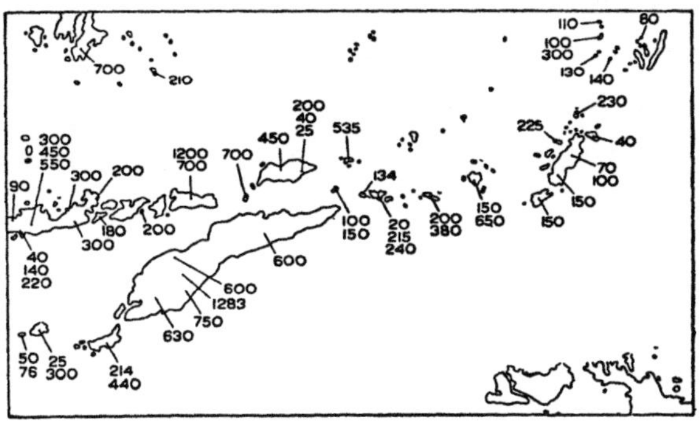

Fig. 34. Elevated reef limestones and terraces in the southern Moluccas. Height in meters.

Kuenen arrived at a similar conclusion in his geomorphological analysis of the bottom relief, and he exhaustively argues that the deep-sea basins originated recently as a result of the subsidence of "continental" (sialic) areas.

This author distinguishes between five groups on morphological grounds (fig. 35). We will discuss four of them, though purposely in a different order. Kuenen's fifth group consists of grabens and other formations, to which we

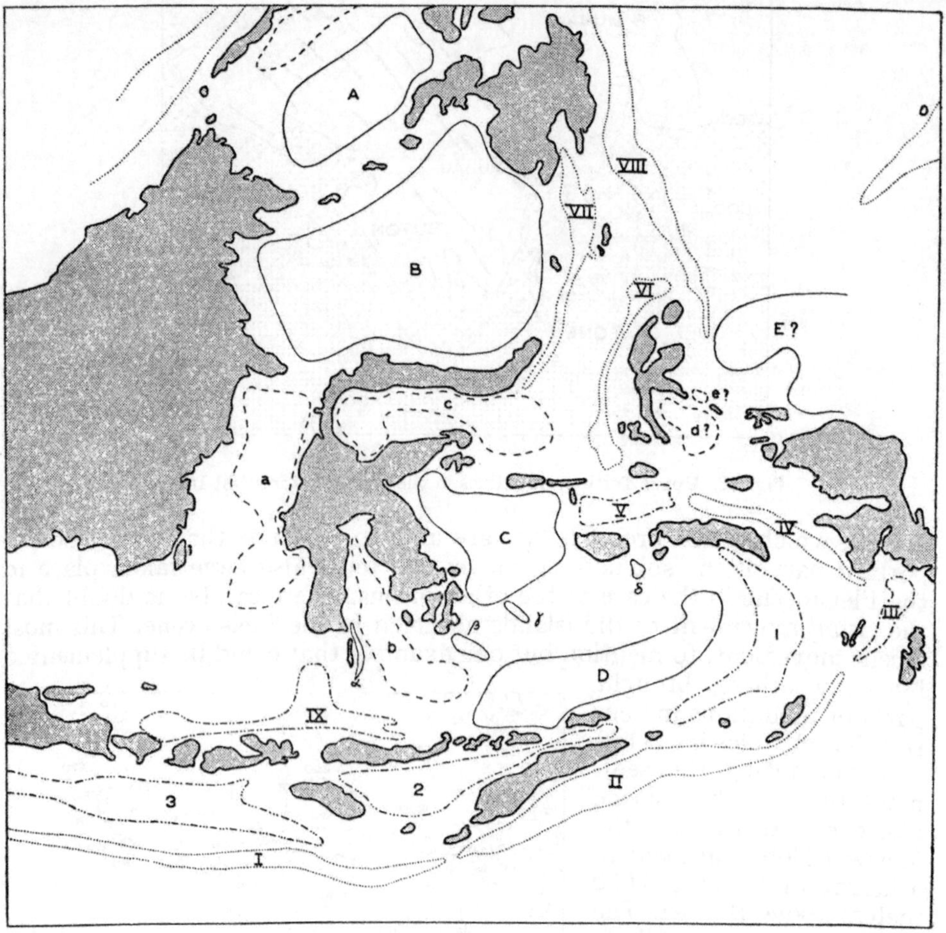

Fig. 35. Types of submarine basins and troughs in the East-Indies (from Ph. H. Kuenen). I—IX marginal deeps; 1–3 troughs of the intramontane type; A—D nuclear basins; a—c discordant basins. I Java trough; II—VII Timor-Ceram and Molucca troughs; VIII Mindanao trough; IX trough of the Flores Sea. 1. Weber deep; 2. Sawu Sea; 3. Java-Mentawei trough. A Sulu Sea, B Celebes Sea, C and D Banda Basins. a Makassar strait; b Gulf of Bone; c Gulf of Tomini.

will not here give our attention. The dubious cases, indicated with a special notation in fig. 35, will likewise not be considered.

Kuenen's fourth group consists of the long furrows on the external side of the zone of miocene folding (fig. 36), i.e. the trough west of Nias and the

Mentawei-Islands, the trough south of Java (which has an occasional depth of 7,000 meters), and the Timor trough. These troughs concur with marginal deeps of the miocene zone of folding as regards the time of their formation, their "tectonic" location and general morphological features.

His third group includes not only considerably shallower formations such as e.g. the Mentawei trough, the trough along the south coast of Java, the Sawu-Sea and the Wetar deep, but also the Weber deep, which has an exceptional depth. They differ morphologically from the preceding group in that they present a less oblong and therefore a more basin-shaped appearance. As regards this group's location, the Mentawei trough and Sawu-Sea have both subsided in the miocene zone of folding (fig. 36), and as the submarine ridge south of Java can be said to belong in a morphological and gravimetrical respect to the miocene belt of folding extending from Mentawei towards Timor (see fig. 36 and 41), the same may probably be

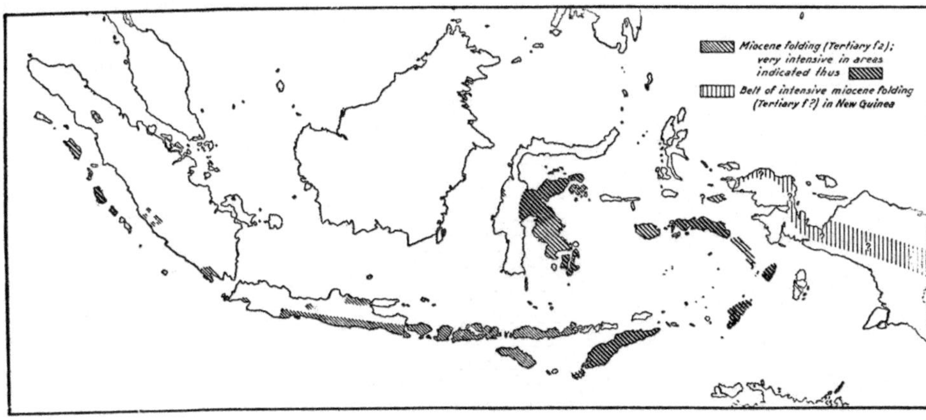

Fig. 36. Areas of miocene folding in the East-Indies.

assumed of the basin situated immediately south of Java. This conclusion also holds good for the Wetar and Weber deeps, since these formations are bounded on one side by the miocene zone which can be followed from Timor via the Tanimber and Kei-Islands towards Ceram and the other side by the submarine continuation of the miocene zone of the Lesser Sunda Islands. These troughs may consequently be said to belong to the intramontane type as regards their situation, morphology and formation.

It is doubtful whether the Flores-Sea ought to be regarded as a marginal deep or as an intramontane trough. Kuenen included it in his fourth group. If this interpretation be correct, we would find a marginal deep on either side of the miocene zone of folding- i.e. the Java-Timor trough on its southern side, and the Flores trough north of it. The Siwalik and Karakum troughs are placed similarly in respect to the Himalayas.

Kuenen's first group is composed of basins with relatively steep sides and flat, horizontal bottoms (the Banda basins, the Celebes-Sea and the Sulu-Sea). Dr. Fr. Weber pointed out that the southern Banda-Sea, or at least the greatest part of it, had lain above sea-level during the Mesozoic. The folded chains group themselves around the actual Banda-Sea, and the latter thus closely resembles such nuclear basins as the Pannonian, Tarim, Ordos and Kara-Sea basins as far as the time of formation and the morphological and tectonic characteristics are concerned.

The Macassar, Bone and Tomini basins make up Kuenen's second group. These formations have a some what shallower flat bottomed cross-section, and a long and irregular shape. The upper-tertiary chains are intersected by the steep coast of the Macassar and Bone basins (fig. 33) [1]).

These basins (at any rate the first two) are similar to those discussed as discordant basins under our Type IV.

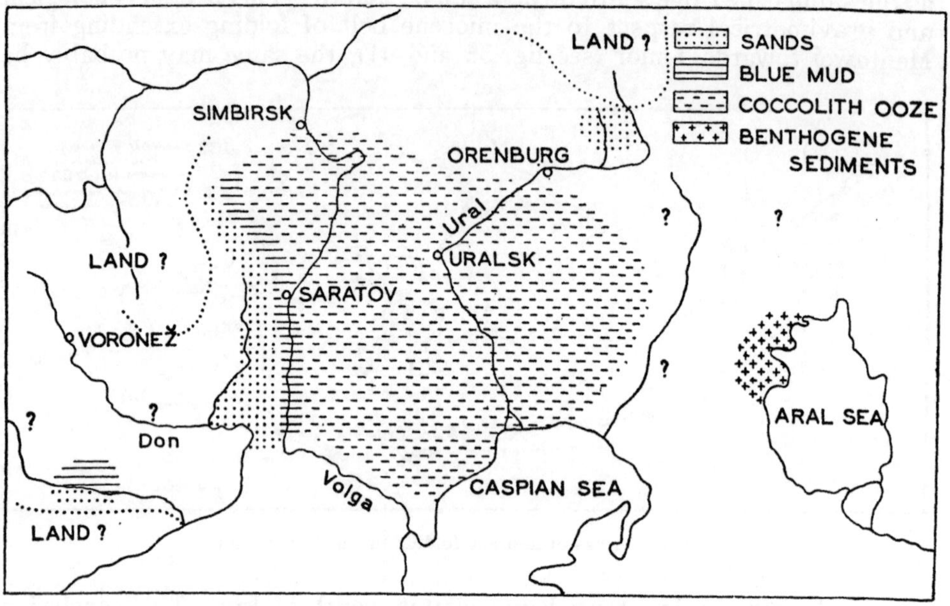

Fig. 37. Distribution of the abyssal coccolith-ooze in the Caspian basin during Maastrichtian times (After S. von Bubnoff).

It seems most remarkable, therefore, that Kuenen's division of deep-sea furrows into four groups should correspond with our own classification of four types of continental basins. We discussed Kuenen's types in a different order because we wished to emphasize the similarity between the results.

One more question remains to be examined, namely whether basins or troughs can be found on the continents, which — like the present East-Indian deep-sea basins — were filled with sediments of bathyal an even abyssal facies. This question can be answered in the affirmative, for the Indian Archipelago itself contains examples of such fossil troughs, e.g. Timor. One or more troughs with bathyal and abyssal sediments existed in Timor during the Mesozoic. A paleogeographic reconstruction of these

[1]) The recent formation of the Gulf of Tomini was pointed out in my publication of the year 1939.

troughs is not possible at the moment, for too little in known of the complicated structure of this island, but the distribution of the facies types leaves no doubt that deep-sea deposits originated in the vicinity of areas of neritic sedimentation. Thus, too, we know that troughs containing bathyal and perhaps even abyssal sediments existed in the complicated geosyncline of the Alps during Mesozoic times. The cherty to siliceous limestone and black shales of the Ouachitas and the Marathon Mountains may possibly be indicative of pelagic conditions and sedimentation in deep water. And, lastly, mention must be made of the Caspian basin (fig. 37), a large portion of which was filled during Maastrichtian times by coccolith and globigerina ooze, and the latter accumulates at a depth of from 2,000 to 3,000 meters according to the latest oceanographic data.

We have enumerated a few examples of fossil deep-sea troughs and basins, and the present deep-sea basins of the East-Indies may consequently safely be classified according to our system. The morphological aspect of an area of subsidence merely depends on the existing relation between the speed of the subsidence and the quantity of detritus supplied. The idiogeosynclines in East-Sumatra and North-Java (fig. 29) were filled with thick deposits because of the huge amount of detritus supplied by the extensive area of Sunda-land, which lay above sea-level at that time. The position of the Mentawei trough is a less favorable one, and this furrow is therefore not filled to a high level. The Weber deep is a very deep trough, as exceptionally few deposits accumulated within it.

We mentioned already that rising movements of the surrounding areas probably accompanied the subsidence of the basins. Nor do conditions in the Indian Archipelago differ in this respect, from those prevailing in the surroundings of the basins on the continents.

Further examples of deep-sea basins will be found in the Appendix, in which reference is made to the basins of Eastern Asia and those of the Mediterranean, West-Indies and southern-Antilles. All these formations are situated inside continental folded chains and loops of continental islands, or immediately alongside of them. Still, the bottom relief of some oceanic areas proper also show basins and ridges of a type similar to that observed on the continents. These will be discussed in Chapter VI.

Deep-sea basins are momentarily very numerous. It is obvious, however, that a comparison with the past has to restrict itself to the continents' present boundaries. Leuchs observed quite rightly that many of the marginal deep-sea troughs lie at such a distance from an area of erosion of any importance (e.g. along the arc-shaped islands in Eastern Asia), that it is difficult to understand how they could be filled by a thick geosynclinal sequence of strata. Another point which deserves attention is that the abyssal sediments which are at present deposited within these troughs contain little or no calcium carbonate. This lack of lime has only been observed in a few exceptional cases in fossil troughs (e.g. in Timor).

Thus, though but a few examples are known of fossil basins which are similar to our present deep-sea basins, we should not forget that these formations are only met with in exceptional cases on the continents. In addition to this the relief of our continents is at the present time particularly high, and not only did the Alpine chains undergo a "rejuvenation" in the form of a rising movement, but even many older chains partook of this process. Moreover, if we stop to consider that the subsidence of a submarine area in many regions is accompanied by a rising movement of the surrounding land, it will at once be clear that vertical movements became very intense within the crust during the latest part of the earth's history, and that it created a strong relief on the continents, and many deep basins below sea-level.

A comparison between the present relief of the floor of the sea and fossil troughs would be misleading, for the continents are the only areas that supply us with data of former periods. No one can tell whether the relief of the ocean's floor had been more accentuated during certain periods than others. Deep-sea troughs cannot be said to have occurred very abundantly periodically; nor can the floor of the deep-sea be said to have flattened out in the intervening time. Still, it would be quite as unfounded to affirm that this certainly was not so. This is one of those typical points which are left open to conjecture.

Chronological relations with other phenomena

The most striking results of an examination of basins and troughs is that the origin and history of these formations are related to certain epochs of folding and mountain-building. This is not surprising in the case of marginal deeps and intramontane troughs, if we consider their situation. The nuclear basins, too, lie inside, or in between certain folded zones, and are surrounded by mountain-chains. It is obvious, therefore, that these basins must be subjected to the influence of processes deeper down in the earth when the latter occur underneath these particular orogenic belts. A problematical point is what happens in the substratum. We might be inclined to associate this question with changes and displacement of magma, but in doing so we would stray too far into the domain of speculation.

One remarkable feature, however, is that even discordant basins show chronological relations with folded belts, though these basins sometimes originate at a great distance from such zones. The subsiding movement of discordant basins, too, begins after a special epoch of folding, and in addition to this, certain phenomena in the history of these basins, such as unconformities, faulting, faint folding and a repeated tendency to subside appear to be chronologically related to specific phases known in folded chains. One thing and another not only point to a deep, internal terrestrial process, but also show that this process has a world-wide activity. Lastly, the greatest part of the basins (as well as of the troughs) form after certain Variscian epochs (see Plate 6). A similar and exceptionally intense formation of basins succeeds certain Alpine phases. Those basins which are connected with Caledonian epochs as far as their origin is concerned, are undoubtedly far less numerous, and only a few can be associated with Lower and Upper-Cimmerian phases (see Table I).

Can it be a mere accident that an identical sequence was arrived at in the case of periods of mountain-building? The given division of basins can no doubt be improved upon. Some readers will perhaps feel inclined to change certain notations. However, I am of the opinion that nothing can alter the remarkable result that the periods of variscian and alpine mountain-building, which were of far greater importance than the caledonian and cimmerian, corresponded with periods in which a considerably larger number of basins were formed (Table I).

It certainly is not a mere accident that the periodicity and intensity of mountain-building is related in point of time to the analogous periodicity and intensity of the formation of basins (Table II).

Epochs of folding, i.e. periods of increasing pressure in the earth's crust, are succeeded by periods of decreasing compression, which also represent periods of mountain-building. As soon as the activity of the periods of intensive compression ceases, the mountains not only begin to rise, but the negative elements, too — the basins — begin to take shape.

As pointed out previously, the shape of some discordant basins is known to be determined by certain structural elements of the basement. Nevertheless so few details are known of these basins that it is in most cases impossible to say what brought about their shape and location. On the other hand, there can be no doubt that they are the result of the whole structure of the foundation and the surrounding areas, and this is probably why subsequent phases of folding are manifested more strongly in one basin than in another. These factors are likewise responsible for the fact that the tendency of a contemporaneous basin to subside is continued for a greater length of time in one basin than in another. The same applies to geosynclines.

Generally speaking, those basins that are found in precambrian Shields are larger than those originating within, or near, a later folded belt.

I hope that the above (including the Appendix, which discusses the various data on which this chapter is based) will have succeeded in showing that the origin of basins and troughs is related to periodical processes in the earth's interior, and particularly that periods of decreasing compression in the earth's crust are characterized both by a rising movement of the folded chains, and the beginning of local tendencies to subside, — in other words the beginning of the formation of new geosynclines and basins.

References

The works on regional geology, which are cited in the bibliography of Chapter II, should also be consulted.

Bärtling, R. *Transgressionen, Regressionen und Faziesverteilung in der Mittleren und Obere Kreide des Beckens von Münster* (Zeitschr. Deutsch. Geolog. Gesellsch. 72, 1920).

Born, A. *Uber jungpaläozoische Kontinentale Geosynclinalen Mitteleuropas* (Abh. Senckenberg Natur f. Gesellsch. Bd. 37, 1921).

Bubnoff, S. Von, *Grundprobleme der Geologie* (1931).

Escher, B. G. *Beschouwingen over het opvullingsmechanisme van diepzee-slenken* (Verh. Geol. Mijnbouwk. Genootschap Geol. Ser. Dl. III, 1916).

Grabau, A. W. *Migration of geosynclines* (Bull. Geol. Soc. China 3, 1924).

Hess, H. H. *Geological interpretation of data collected on cruise of U. S. S. Barracuda in the West Indies* (Preliminary Report. Transact. Americ. Geophysical Union 1937).

HESS, H. H. *A new bathymetric chart of the Caribbean area* (Advanced report of the commission on continental and oceanic structure 1939).

HOLTEDAHL, O. *Geologische Karte der Arktis mit angrenzenden Gebieten* („Arktis" 3, 1930).

KAHRS, E. *Zur Palaeogeographie der Oberkreide in Rheinland und Westphalen* (N. Jahrb. Geol. u. Pal. Beil. Bnd. 58, 1927).

KUENEN, PH. H. *Geological interpretation of the bathymetrical results* (The Snellius Expedition, Vol. 5, part. 1, 1935).

LAWSON, A. C. *Insular arcs, foredeeps, etc.* (Bull. Geol. Soc. America 43, 1932).

LEMOINE, P. *Considerations sur la structure d'ensemble du bassin de Paris* (Livre Jubilaire Centenaire de la Soc. Geol. de France, Tome II, 1930).

LEMOINE, P.; HUMERY, R. et SOYER, R. *Les forages profonds du bassin de Paris. La nappe artésienne des sables verts* (Mem. Muséum Nat. Hist. Nat. 1939).

LEUCHS, K. *Tiefseegräben und Geosynclinalen* (N. Jahrb. Min. etc. Beil. Bnd. 58B, 1927).

MOLENGRAAFF, G. A. F. *On recent crustal movements in the island of Timor and their bearing on the geological history of the East Indian Archipelago* (Proceed. Kon. Acad. v. Wetensch. Amsterdam 1912).

PRUVOST, P. *Sédimentation et subsidence* (Livre Jubilaire, Centenaire de la Soc. Geol. de France, Tome II, 1930).

REED, R. D. *Southern California as a structural type* (Bull. Americ. Assoc. of Petrol. Geol. 21 1937).

SCHUCHERT, CH. *Sites and nature of the North American Geosynclines* (Bull. Geol. Soc. America 34, 1923).

SCUPIN, H. *Palaeogeographie* (1940).

STEVENS, CH. *Le relief de la Belgique* (Mém. de l'Instit. Geolog. de Louvain. XII. 1938).

STILLE, H. *Alte und junge Saumtiefen* (Nachr. K. Gesellschaft der Wiss. Göttingen. Math. Phys. Klasse 1919).

STILLE, H. *Die Oberkarbonisch-Altdyadischen Sedimentationsraüme Mittel-Europas in ihrer Abhängigheit von der Variscischen Tektonik* (Congr. de Stratigr. Carbonifère. Heerlen 1927, Ed. 1928)

TEICHERT, C. *The Mesozoic transgressions in Western Australia* (The Australian Journal of Science, Vol. 2, 1939).

UMBGROVE, J. H. F. *Verschillende typen van tertiaire geosynclinalen in den Indischen Archipel* (Leidsche Geol. Mededeelingen VI, 1933).

UMBGROVE, J. H. F. *De pretertiaire historie van den Indischen Archipel* (Leidsche Geol. Mededeelingen, VII, 1935).

UMBGROVE, J. H. F. *On the time of origin of the submarine relief in the East Indies* (C. R. Congr. Internation. de Geographie, Amsterdam 1938).

UMBGROVE, J. H. F. *Atolls and barrier reefs of the Togian islands, North Celebes* (Leidsche Geolog. Mededeelingen. XI. 1939).

VEATCH, A. C. *Evolution of the Congo Basin* (Geolog. Soc. of America, Memoir. 3. 1935).

WATERSCHOOT VAN DER GRACHT, W. A. J. M. VAN, *Lateral movements on the Alpine Foreland of North Western Europe* (Proceed. Kon. Acad. v. Wet. Amsterdam, vol. 41, 1938, p. 195—214)

Chapter IV

CRUST AND SUBSTRATUM

"Pas de synthèse tectonique sans vision d'un continu à trois dimensions en train de se déformer". (E. Argand)

Introduction

Chapter I gave a brief outline of those rocks which can probably be said to build up the earth's crust. This important question will now have to be examined somewhat more closely. It should first be noted, however, that opinions differ considerably as regards the composition of the terrestrial crust though all agree on a few fundamental points.

An imaginary cross-section of a continent would probably present more or less the following aspect (fig. 38). On top we would find a veneer of sedimentary strata, totalling as much as 15 km in the geosynclines, or possibly even more, but thinning out to zero at points where their foundation is exposed. Beneath this veneer would stretch a zone of acid rocks, consisting mostly of granite and gneiss, and descending to a depth from 10 to 30 km below the cover of sediments. The thickness of the crust in the continental regions is thought to vary between 30 and 50 km. Gutenberg's latest review shows that comparatively smaller figures apply to such regions, for instance, as Western Europe, New-Zealand and the north-eastern part of Japan. The largest values found are in the Sierra Nevada, California (40—65 km) and the Alps (50 km). The lower side of the crust is known to form an important primary discontinuity — "the Mohorovičić discontinuity" There is no one to-day who doubts that the lower part of the continental crust is composed of basic rocks with an occasional thickness of 15—30 km.

Opinions disagree as regards the nature of these basic rocks, which are piezo-gabbro or olivine basalt according to some authors (Daly, Kennedy), and peridotite according to others (Holmes, Hess, Eskola and others). The most important point, however, as far as we are concerned, is that this lower part of the crystalline crust is now universally assumed to consist of basic to ultrabasic rocks, the so-called sima. This sima's chemical composition differs but slightly from that of the substratum upon which rests the crust. The lower layer of the crust is indeed regarded as a layer of crystalline sima. Some writers, including Daly, assume that amorphous sima — the "vitreous substratum", an overheated and very viscous basaltic liquid under high pressure — follows beneath the Mohorovičić discontinuity.

To avoid any misunderstanding, I want to emphasize that the terms "sima" and "sial" (which were introduced by Suess) are used in the petrographical sense, and not in Wegener's physical sense, such as rigid and elastic sial and viscous sima.

Some authors are of the opinion that one, or possibly more than one layer, with a thickness of about 25 km, extend between these two parts of the earth's crust. There are many conflicting views as regards the composition of this intermediate layer, some assuming it to consist of granite or diorite, and others of granodiorite or granulite, and even of amphibolite, olivine-free basalt or gabbro.

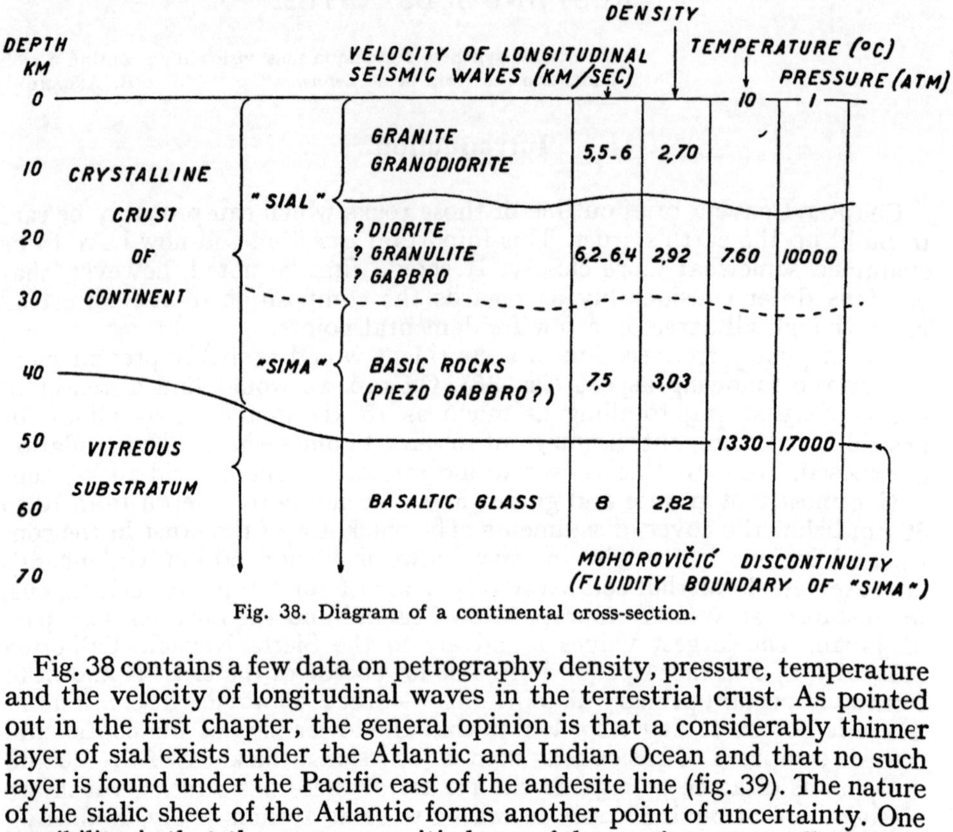

Fig. 38. Diagram of a continental cross-section.

Fig. 38 contains a few data on petrography, density, pressure, temperature and the velocity of longitudinal waves in the terrestrial crust. As pointed out in the first chapter, the general opinion is that a considerably thinner layer of sial exists under the Atlantic and Indian Ocean and that no such layer is found under the Pacific east of the andesite line (fig. 39). The nature of the sialic sheet of the Atlantic forms another point of uncertainty. One possibility is that the upper, granitic layer of the continents as well as one or more of the intermediate layers, are thinning out towards the oceans. Another supposition is that one of these layers is entirely absent in the Atlantic sector [1]). We will return to this question in Chapter VI. Though the floor

[1]) Vening Meinesz pointed out that the gravimetric profiles show a sudden increase of the positive anomalies (of 30—100 milligals) when passing from the shelf to deep water. After a discussion of the available seismic and gravimetric data, he comes to the conclusion that the most probable explanation is.... "to assume the granitic layer only to be present in great thickness in the continents and to end rather suddenly at the edge of the shelf or, if

of the Pacific differs in volcanologic, petrographic, magnetic and seismic respect from a continent, the same cannot be said of the floors of the Atlantic and Indian Ocean and the surrounding continents [1]).

The crystalline crust of the Atlantic, where seismic data (together with the evidence found in the granitic inclusions observed in the lavas of some volcanic islands [2]) have revealed the presence of a thin layer of sial, must be quite as thick and possibly even thicker than that of a continental sector. Under the Pacific, where a sialic layer appears to be absent, the rigid crust, consisting entirely of crystalline sima, is thicker than anywhere else (fig. 39). Daly's view is based on the assumption that the thickness of

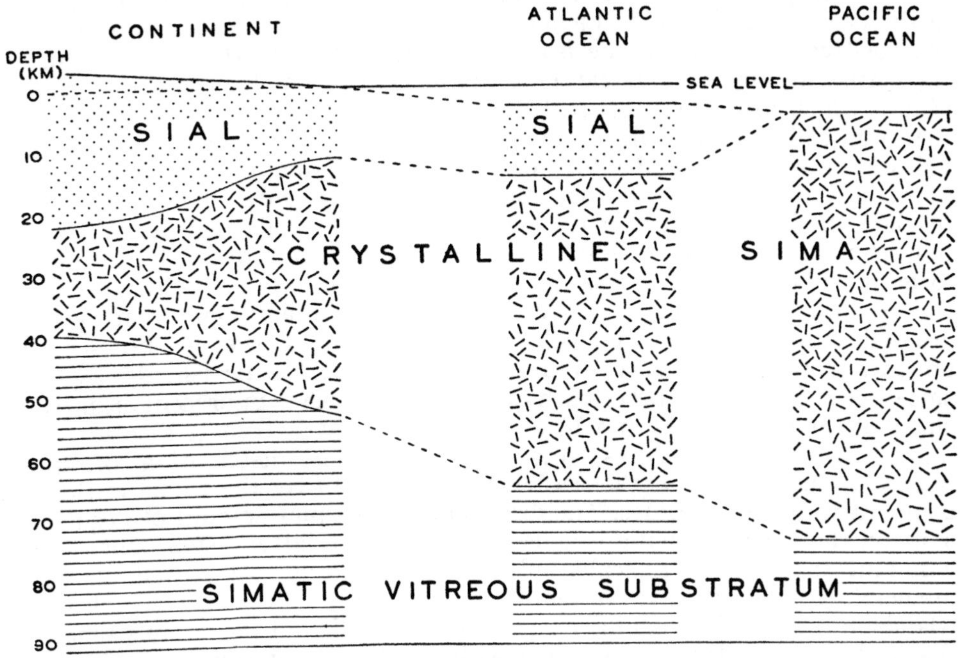

Fig. 39. Schematic sections across the earth's crust.

the crust depends upon the production of heat by radio-active minerals, the sial being considered to be more radio-active than the basic rocks of the sima. The terrestrial crust will consequently be thinner under a sialic continent, when in a state of thermal equilibrium, than under an ocean,

---

present in some parts of the oceans to be thin there. The next layer seems to continue under the oceans, perhaps somewhat thinner under the deepest parts of the Atlantic and Indian oceans and certainly thinner under the central part of the Pacific east of the andesite line; it may even be absent in this last area but in this case we must assume a still deeper crustal layer to be present there of a smaller density than the layer below it. The principal part of the gravimetric shelf- profiles as well as of the seismic data could thus be explained".

1) Gutenberg 1939, p. 302 — 320.

2) Ascension e.g., including probably Tristan da Cunha. Plutonic rocks, together with sediments of varying ages and metamorphous rocks are also known to occur in some of the Cape Verde Islands and the Canaries.

where the sial is either thin or non-existent, and the solid crust will have to be even thinner under the additional quantity of sial of the "root" of a folded mountain-chain [1]). The lower side of the crystalline crust should at the same time be regarded as the upper limit of melting of the sima, since the temperature at this depth is equal to the melting point of the rocks under the prevailing temperature.

Rittmann's model of the structure of the earth's crust was based on similar arguments as those of Daly, and their view was lately confirmed by investigations of an entirely different kind by Vening Meinesz. A study of gravity fields over the Hawaiian Archipelago and the Madeira area has shown, amongst other things, that there cannot be any important difference in the crustal rigidity and thickness beneath both these areas [2]), and that the positive anomalies of the isostasy strongly indicate the existence of a rigid crust, acting as an elastic sheet that is bent by the load of the volcanic islands which formed on top of it.

It was claimed by Molengraaff in 1916 that a subsiding movement of volcanic islands in the Pacific would be brought about by slow isostatic downward movements, the subsidence continuing until the islands had reached the level of the simatic bottom, which he regarded, as Wegener did, as a viscous liquid. Molengraaff thus sought to explain the ultimate formation of thick coral reefs, and the creation, in conformance with Darwin's theory, of a barrier reef, and, finally, an atoll, from a fringing reef by the gradual subsidence of the volcanic foundation. However, no strong downward movement is apparent in many of the Pacific islands, as has since been shown by observations along their coasts. Subrecent negative shifts of the shoreline with a height of approximately 6 meters occur in a number of these localities, and the world-wide distribution of this phenomenon proves that the sea-level has lowered over the whole world in comparatively recent times. The discovery of an equal amount of negative shift on the simatic Pacific islands shows that the latter's subsidence (i.e. assuming that they were really subsiding) has at any rate ceased. Vening Meinesz' geophysical investigations show that the subsidence of volcanic islands in the Pacific is confined to the down-bending of the ocean's simatic floor, and that the latter acts as an elastic sheet, and not as a viscous liquid, as Wegener and Molengraaff supposed.

The conception of the earth's crust at which we arrived, above, explains several striking volcanological and petrographical phenomena. Only basic magma can reach the surface where the sialic layer is lacking, and this material — through differentiation — can furnish trachyte to phonolite (the so-called atlantic rocks), but no quartz-bearing acid rocks, and therefore certainly no granite. If the basic magma of the simatic substratum suddenly invades a continental crust by a so-called abyssolithic injection into faults, the only material to extrude at the surface will again be basic material [3]).

---

1) Daly, 1938, p. 35.

2) The figures assumed by Vening Meinesz as regards the thickness of the crust (25 to 45 km) are much lower than those of Daly mentioned above.

3) The possible genesis of the "mediterranean" rocks will not be dealt with here, for the various hypotheses on their origin (whether they be Rittmann's, Daly's, or those of other authors)

Tectonic and magmatic cycles

We now return once more to the question of a geological cycle, a diagrammatic view of which will be found in fig. 12. Should the terrestrial crust be drawn into this figure, it would become clear that a thick downward bulge of acid basement rocks must have penetrated the underlying, denser basic rocks during the process of crustal folding. A root of lighter rocks must then have formed under the folded geosyncline. Hess illustrated this in the case of the Alps (see fig. 40). It ought to be possible to determine the existence of such a root by means of gravimetrical observations, as the same would constitute an anomaly of the isostatic equilibrium.

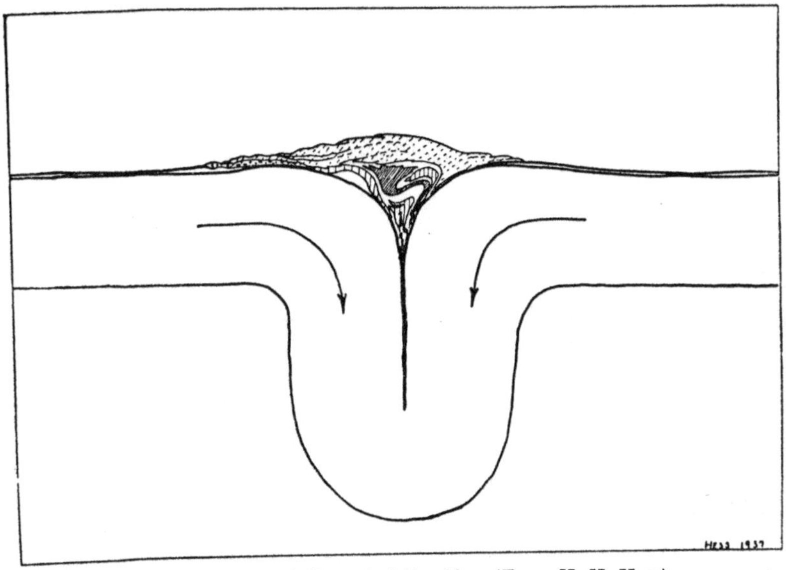

Fig. 40. The sialic root of the Alps. (From H. H. Hess).

This sounds obvious, but, remarkably enough, the isostatic anomalies were in fact observed first, and from these was derived the existence of a sialic root. There can be no doubt, as Griggs remarked, that "one of the greatest contributions to the understanding of tectonics during the twentieth century has been Vening Meinesz' discovery of the great bands of gravity deficiency in the East and West-Indies" [1]). Vening Meinesz discovered abnormally large isostatic anomalies during his gravity expeditions in the East-Indies in 1929 and 1930. One of the principal features of these deviations is a narrow belt of strong negative anomalies (fig. 41), which was found to continue for about 5,000 miles. Anomalies are expressed in milligal,

cannot be said to have a direct bearing on the present subject.

A clear table of petrogenese appeared in one of Rittmann's publications (Geolog. Rundschau, 30, 1939, p. 603), and a map on the distribution of the recent volcanic rocks will be found in the book which this author published in 1936.

1) Griggs, 1939, p. 614—616.

i.e. the third decimal — or, roughly, millionth parts — of gravity [1]). Anomalies of more than 50 milligal are but seldom observed, but this strip shows anomalies of more than 100, and occasionally of even more than 200 milligal. An analogous band was discovered in the West-Indies by the expeditions of Hess, Browne, Vening Meinesz, Hoskinson and Ewing (fig. 41). These remarkable strips of negative anomalies are undoubtedly due to an abnormal distribution of crustal material beneath the strip. Vening Meinesz' explanation — the only one that seems to cover all the facts — is that a great

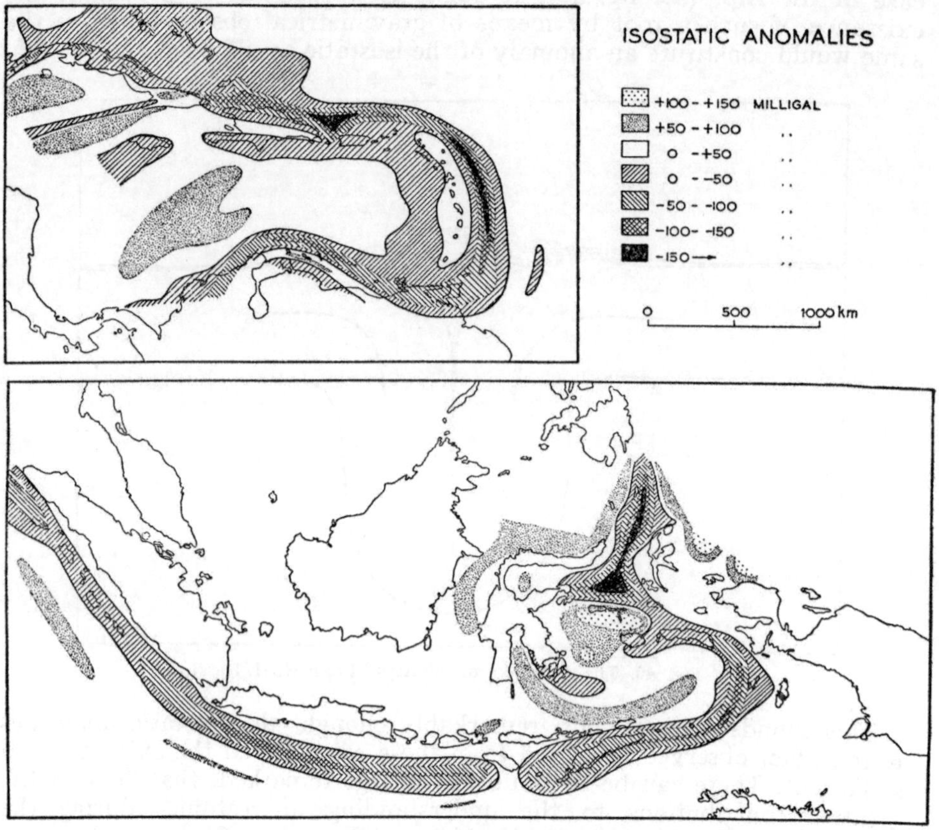

Fig. 41. Isostatic anomalies in the West-Indies (after H. H. Hess) and the East-Indies (after F. A. Vening Meinesz).

isoclinal protuberance of the crust was downfolded into the substratum. This lead him to suppose "that the crust under the action of great horizontal stresses is buckling inwards, and that the folding and overthrusting of the surface layers, as found by the geologists, are an accompanying feature of this great phenomenon" [2]). It would be impossible for us, at this point, to

1) Normal gravity amounts to 987.049 gal at the equator, and to a thousand times as many milligal.

2) Vening Meinesz 1933, p. 372.

enter into an examination of the details of either the accompanying strips of positive anomalies, or their relation to the structural history — and seismic and geomorphological features — of the explored areas. These questions are dealt with extensively in the publications by Vening Meinesz, Hess, Griggs and Kuenen appearing in the list of references at the end of the chapter.

We are at present particularly concerned with those phenomena in

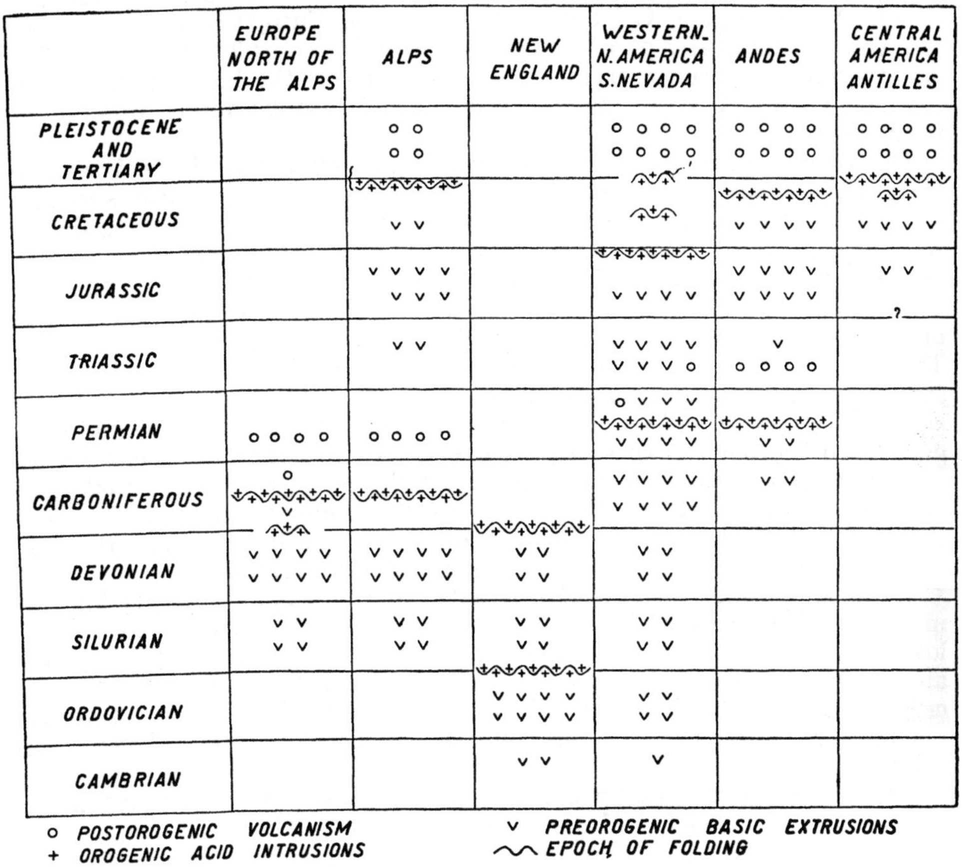

Fig. 42. Relations between magmatic phenomena and epochs of folding in different regions. Compare with Table II (After H. Stille).

general which may be said to be associated with a subsiding and subsequently buckling crust. One characteristic phenomenon is that basic to ultrabasic rocks are found in the axis of a geosyncline, but that no acid rocks are observed at this stage of the development. Conversely, acid intrusions enter the geosynclinal belt during and after its folding. Thus an outpouring of large flows of lava is known to have occurred in the jurassic Fransiscan beds of the Sierra Nevada geosyncline, and these flows are actually found as intercalations between the strata. Diabase, gabbro and ultrabasic intrusions (serpentine), and, lastly, volcanic products, the latter mingling with the marine sediments, accompanied these outflows. Intensive folding ensued, during which the numerous batholithic intrusions of granodiorite, quartzdiorite and quartzmonzonite were emplaced [1]). Kossmat described a similar sequence in the Alpine, Variscian, Caledonian and older folded belts of Europe and in the Australian Caledonides. Stille grouped some European and American examples systematically in a table (fig. 42). The Alpine, Variscian, and Caledonian Mountains of Europe had already been grouped schematically at a previous date by both Stille and Lotze. This last diagram has been reproduced in Table II (see the last column at the extreme right). A great many examples will also be found in regional treatises [2]).

This remarkable sequence of magmatic rocks can also be derived without the aid of strained hypothetical constructions from the previously sketched structure of the earth's crust if we assume that a zone was first subjected to geosynclinal subsidence, and that it subsequently buckled as Vening Meinesz has suggested. The following considerations require but one assumption, i.e. that an upper layer of acid rocks rests on top of a layer of basic rocks. The intermediate layer will be left out to make things simpler. The block-diagrams of fig. 43 show that the crystalline crust consists of a sialic layer of approximately 20 km and this is seen to lie on a basic layer of equal thickness. It will be obvious that the ensuing reflections would hold good even if other thicknesses were concerned. In the meantime, I would like to make it clear that fig. 43—45 should be regarded as rough and entirely schematic sketches of the events. Moreover, anyone who has attempted to make such a drawing will have realized for himself that many dubious and difficult points arise in the course of such an undertaking.

When an area of subsidence originates in the earth's crust, sediments may accumulate at the surface in abnormal "geosynclinal" thickness. The lower part of the crystalline crust enters an area of higher temperature and pressure and subsides to the fluidity boundary. If that part of the crust were to melt, it would only be able to furnish basic to ultrabasic magma similar to the vitreous substratum already present beneath it (block I).

The strength of the crust will be exceeded when the downward movement reaches a certain level, and disruption will follow. The basic magma will then immediately (since it is subjected to high pressure) fill the fissures and faults in the crust. As the bending of the crust is strongest in the axis of the geosyncline, basic magma in this part will have the greatest chance of reaching the surface along faults. Moreover, that portion of the crystalline

[1]) Waters and Hedberg, Reg. Geol. der Erde, 1939.

[2]) With the aid of data on radioactive determinations of age, Kuenen recently showed that, pending confirmatory data, the relatively short periods of intrusion, which were separated by longer periods of quiescence, can similarly be determined in Precambrian times, and concluded that "the ultimate cause must lie below the crust and must be world-embracing". (Kuenen 1941, p. 337).

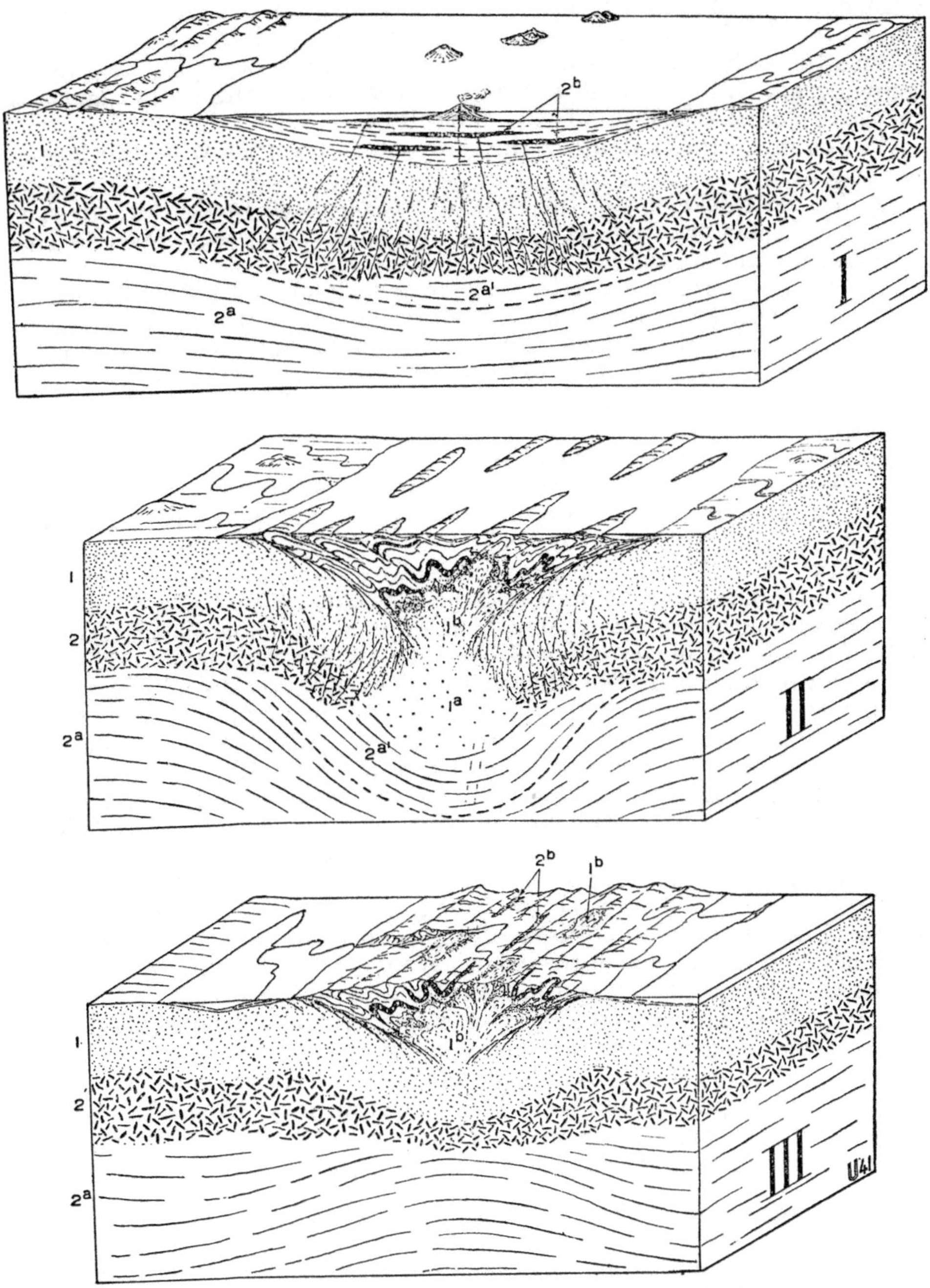

Fig. 43. Schematic and tentative block diagrams illustrating the relation between tectonic and magmatic cycles. 1 and 2 crystalline crust; 1 sialic layer; 1a molten part of 1; 1b acid batholithic intrusions and metamorphic aureole; 2 basic layer; 2a basic vitreous substratum; 2a¹ molten part of 2; 2b basic abyssolithic injections, extrusions and volcanism.

crust may be considered to have thinned as a result of subsidence, which caused the fluidity boundary to shift to a higher level (block I).

Once the sialic crust has been bent so far that it collapses, one of the consequences will be that it will buckle inwards into the deeper layers of the earth's crust (block II). The result will be:

(1) Folding of the contents of the geosyncline, together with the latter's basic rocks.

(2) A local thickening of the sialic crust, resulting in the formation of a "root" of sialic material, penetrating into an area of progressively higher temperature and pressure.

(3) When this root's material, begins to melt and invade the surface strata as batholithic intrusions, plutonism and volcanism cannot but be strongly acid ("pacific") in character, since the batholiths assimilate more and more sialic material from the root, and eventually from the sediments of the already folded geosyncline (acidification). The ten characteristics of batholiths, and the explanation which Daly gave of them [1]), fit the diagrammatic explanation given here.

(4) As soon as the compressive forces in the sialic crust decrease, the "root", in order to resettle into isostatic equilibrium, will have to have a tendency to rise. This effect is heightened by the expansion of the sialic root, which, trough fusing, grows specifically lighter. This explains how a mountain-chain is formed from a folded geosyncline. More and more deeply situated batholiths are exposed by denudation at this stage (block III).

(5) The upward movement will continue till the isostatic equilibrium has been re-established. The sima will only be able to crystallize beneath the upper sialic layer to a thickness corresponding approximately with that of the sialic layer after part of the sialic root, situated at an abnormal depth, has spread out and disappeared (i.e. the crystallized sima will be thinner as the sial grows thicker) [2]). As an accompanying phenomenon of the rising zone, sima will flow in from the neighboring strips of the earth's crust. This process may perhaps explain why a depression or "marginal deep" originates on one side, or both sides of a mountain belt.

As pointed out in Chapter II, the "zonal migration" of geosynclines can be traced distinctly in the history of some continents from the Cambrian up to our own time. Though the exact nature of events in the substratum cannot be determined, the subsidence of a geosyncline may be stated to be the result of internal terrestrial processes, whereas the sedimentation may be described as a secundary, external effect.

Fig. 42 and Table II also include postorogenic volcanism, which has been indicated with separate notations. This volcanism may perhaps be partly due to differentiation of the orogenic batholiths, but it is also partly related to the dome-shaped elevations which will be discussed presently.

Zonal migration of geosynclines. Continents and ocean-floors

The oldest cores of the continents are known to consist of intensely folded

---

1) Daly, 1938, p. 169—173.

2) Daly, 1938, p. 190, fig. 151.

and migmatized belts. None of the oldest areas observed so far can be regarded as a primary, i.e. an unfolded fragment of the earth's original sialic crust. That part of the drawing in fig. 43 which represents the seemingly simple structure of a geosyncline's basement, or of an area of folding, can undoubtedly be stated to be very complicated, for it passed through one or more geological and magmatic cycles. (fig. 44 therefore acts as a supplement to fig. 43, II). This fact, if born in mind, might induce us to suppose that the continents might have originated as a result of periodical folding, culminating in a consecutive thickening of a relatively thin sialic layer originally enveloping the entire earth. We then feel inclined to enquire

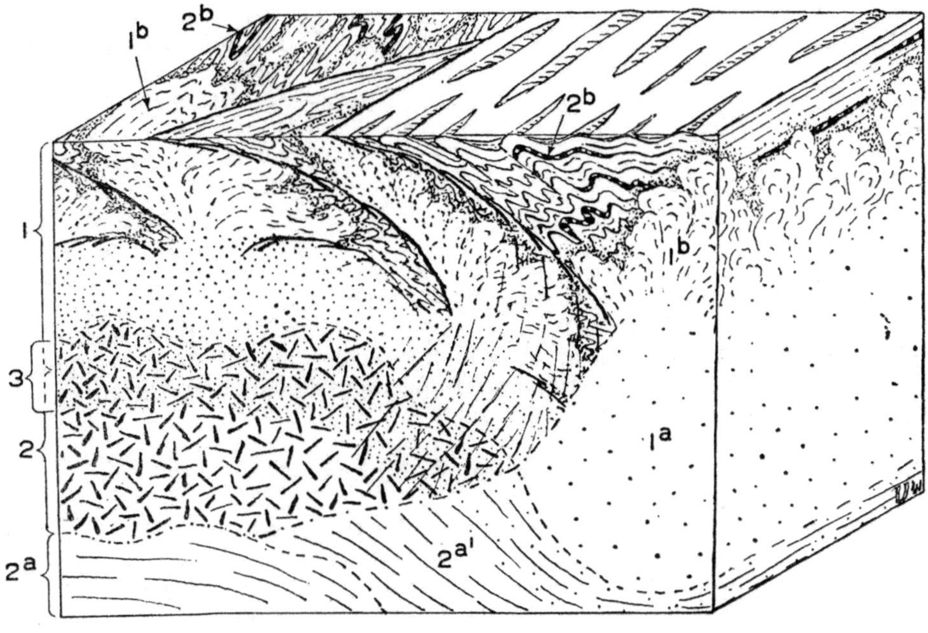

Fig. 44. Tentative illustration of four subsequent tectonic and magmatic cycles in a continental area, completing the left part of block II in fig. 43. For the figures' explanation see fig. 43 and 45.

whether two problems cannot be explained simultaneously, viz. (1) the formation of continental blocks (in which geosynclines and folded mountain-chains originated subsequently according to the complicated patterns discussed in Chapter II), (2) the formation of sial-free parts of the earth, such as the floor of the Pacific (fig. 68). These speculative reflections differ to a considerable extent from the hypotheses of Fisher, Wegener, Schwinner, Escher and others mentioned in Chapter VI. For, unlike these authors, I am of opinion that an upper and acid sialic layer which enveloped the whole earth during a very early stage of its history (following the moon's separation from the earth) was indeed stretched over a basic lower layer. In the beginning the material of both layers must have originated from a primordial melt containing the combined elements of the diverse now contrasting crustal and deeper shells.

These entirely hypothetical considerations lead directly to the important problem of the origin of continents and oceans, but this question will be dealt with in Chapter VI. However, I would like to draw attention to an important deduction. If we assume the primordial crust to consist of two layers — an upper acid and a lower basic layer, indicated as 1 and 2 respectively in fig. 45 — la would represent the acid melt formed as a result of buckling of the crust. This liquid, however, will probably not only move upwards in the shape of batholithic intrusions (1b), but will also flow out laterally in the direction of the arrows. If the molten sial (1a) and sima (2a and 2a¹) recrystallize during a later stage, i.e. during the rising of the root, a zone containing a mixture of la and 2a will have to form in the crust. This may help to explain the formation of a layer such as 3, which is intermediate to the layers 1 and 2 both as regards its position and its petrographical composition (zone 3 has been indicated similarly in fig. 44).

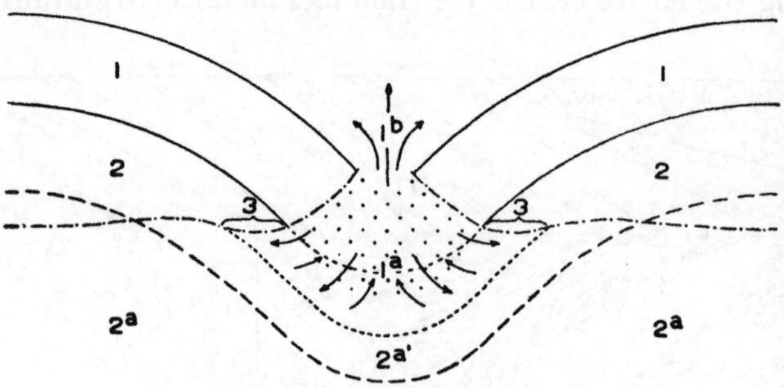

Fig. 45. The possible origin of an intermediate continental layer (3). For an explanation of other figures see fig. 43 and 44.

Volcanism in basins

The foundation of a subsiding basin, like the down-bending crust of a geosyncline, will be affected by fracturing and become injected with magma. The magmatic products intrude as sills, and flow out at the surface as lava flows, or mingle with the sediments as clasmatic products. This can be observed in many basins, and it would certainly be worth while to investigate to what depths the bottoms of geosynclines and basins subsided before the first series of basic rocks made their appearance. This would provide us with valuable data as regards the possible mechanical process in the earth's crust. Repeated reference is made in the descriptions of these basins to basalt, gabbro, diabase, melaphyr and other basic rocks, but no acid rocks have ever been observed so far as I know. It need hardly be said that those basins which are closely connected with zones of folding (the basins of the intramontane type e.g.) should be considered *cum grano salis*, for pacific rocks in their immediate vicinity invaded the crust before and at times even during their formation. It is certainly noteworthy that the marginal deep which originates subsequently to a mountain's principal epoch of folding (cf. Chapter III) should, generally speaking, seem devoid of volcanism. This negative characteristic must have some connection with the deep crustal buckle which was not formed until a short while before and along which the basins and troughs are arranged. A possible explanation may be that a marginal deep, which develops during periods of decreasing compression, has a very short existence, while the major geosynclines and numerous basins have a long history, continuing into one or even several subsequent periods of strong increasing compression of the earth's crust.

Dome-shaped elevations and rift-zones

Cloos recently published an extensive study of the opposite phenomenon — the dome-shaped elevations, and attempted to show that fault systems

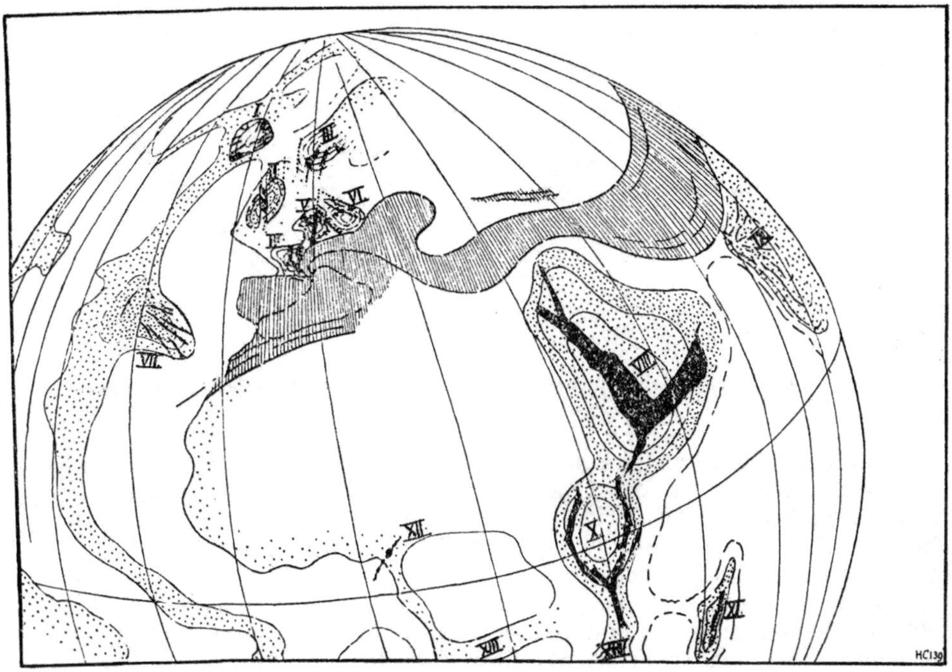

Fig. 46. Dome-shaped elevations, with rift-valleys and volcanism (After H. Cloos). I. Ice-land, II. Great Britain and Ireland, III Graben and faults of Oslo etc., IV Plateau Central of France with the Limagne-Graben, V Rhine-shield and graben, VI Bohemia—Saxonia, VII Azores, VIII Arabia—Nubia, IX India, X East-Africa, XI Madagascar.

and rift-valleys accompanied by volcanism are as a matter of fact the direct result of (and locally dependent upon) slowly updoming nuclei. Examples of such nuclei are the Vosges and the Black Mountains with the Rhine graben, the Plateau Central with the Limagne graben, the African-Arabian Shield with the riftzone of East-Africa, the Red Sea, etc. (fig. 46). Cloos is of opinion [1]) that the first dome-shaped elevation of the Rhine Shield probably originated in the Upper-Paleozoic, and that it was chronologically related

1) Cloos, 1939, p. 462.

to variscian mountain-building, whereas the present phenomena would be due to a much more intensive process of elevation in the Tertiary, and this would have been chronologically related to Alpine epochs of movement. Deep tension-faults and rifts affected the vault's upper layers as a result of this updoming (fig. 47). An older updoming tendency had probably also preceded the formation of the Arabian-African Shield and their rift-systems [1]).

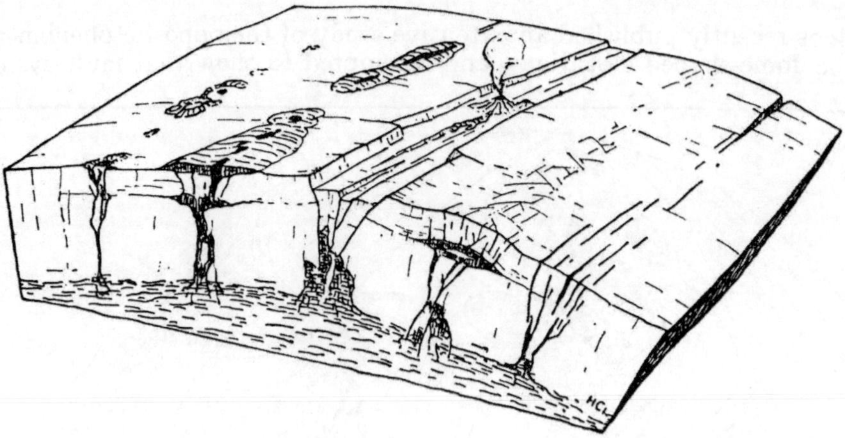

Fig. 47. Block diagram of a dome-shaped elevation with faults graben and volcanism (From H. Cloos).

This induced Cloos to take an entirely different view of the subject than van Waterschoot van der Gracht advocated in 1938, viz. that the rift-systems of Europe and Africa represent the first stage of the disruption of blocks, which — as Wegener postulated — are beginning to drift towards the Atlantic or Indian Oceans.

My own opinion, however, is that the updoming is brought about by a similar phenomenon as the depression of the basins, viz. by a displacement of magma in the substratum. We have even fewer data to go by, here, than in other cases, but the dome-shaped elevations, too, would appear to give indications of their own periodical occurrence and chronological relation to the tectonic and magmatic cycles discussed previously. A great deal more will have to be known of the earth's history, and especially of its precambrian history, before an explanation can be given why the crust arches up in one area, while basin-shaped depressions are formed in another.

Summary

The conception of the structure of the earth's crust, as advocated above, makes it clear that the presence of certain magmatic rocks must (assuming that a period of geosynclinal subsidence is followed by crustal buckling, as

[1]) Cloos 1939, p. 442.

postulated by Vening Meinesz) be related to the phenomena of geosynclinal subsidence, folding and mountain-building. Thus, it is obvious, for instance, that only basic to ultrabasic rocks can occur in a geosyncline, and that intrusions of acid batholiths take place after the crust has buckled and the contents of the geosyncline been folded. All evidence shows that periods of increasing compression of the earth's crust alternate with periods of decreasing compression. During this last stage the sialic root (which was brought about by buckling of the crust) is given an opportunity to rise until the isostatic equilibrium has been restored. A mountain-chain, in a geographical sense, then comes into existence. This process has thickened the sialic layer in the meantime, and the continents may consequently have originated as a result of periodical folding and thickening of an originally thinner sialic layer enveloping the whole earth.

The accumulation of sediments in a geosyncline is a secundary and external effect of a subsiding crustal furrow, and this phenomenon, together with the alternating periods of increasing and decreasing compression, the migration of geosynclines, and (probably) the formation of dome-shaped elevations with their faults and rift systems, must be attributed to some common and world-embracing, deep-seated cause, for all the phenomena are related chronologically.

References

BUBNOFF, S. VON, *Grundprobleme der Geologie* (1931).
BUCHER, W. H. *The deformation of the Earth's Crust* (Princeton Univ. Press 1933).
CLOOS, H. *Hebung — Spaltung — Vulkanismus* (Geol. Rundschau 30, 3A, 1939).
DALY, R. A. *Architecture of the Earth* (1938).
DALY, R. A. *The roots of Volcanoes* (Trans. Amer. Geophysic. Union. Report General Assembly 1938).
ESKOLA, P. *Wie ist die Anordnung der äusseren Erdsphären nach der Dichte Zustande gekommen?* (Geol. Rundschau 27, 1936).
GRIGGS, D. *A theory of mountain-building* (Americ. Journ. of Sci. 237, 1399.
GUTENBERG, B. *Internal constitution of the Earth* (Physics of the Earth VII, 1939).
HESS, H H.. *Island arcs, gravity anomalies and serpentine intrusions* (Proc. 17th Intern. Geol. Congress, Moscou 1937).
HESS, H. H. *Primary peridotite magma* (Americ. Journ. Sci. 35, 1938).
HESS, H. H. *Gravity anomalies and island arc structure with particular reference to the West Indies* (Proc. Americ. Philos. Soc. 79, 1938).
HOLMES, A. *Contributions to the theory of magmatic cycles* (Geol. Magaz. 62, 1925; 63, 1926).
HOLMES, A. *Radioactivity and earth movements* (Trans. Geol. Soc. Glascow 18, 1928—'29).
HOLMES, A. *The origin of igneous rocks* (Geol. Magaz. 69, 1932).
KENNEDY, W. Q. and ANDERSON, E. M. *Crustal layers and the origin of magmas* (Bulletin volcanologique II, 3, 1938).
KOSSMAT, F. *Paläogeographie und Tektonik* (1936).
KOSSMAT, F. *Der ophiolitische Magmagürtel in den Kettengebirgen der Mediterranen Systems* (Sitzungsber. Preuss. Akad. d. Wissensch. 24, 1937).
KUENEN, PH. H. *The negative isostatic anomalies in the East Indies* (Leidsche Geol. Mededeelingen VIII, 1936).
KUENEN, PH. H. *Major geological cycles* (Proc. Nederl. Akad. v. Wetensch. Amsterdam XLIV, 1941).
MOLENGRAAFF, G. A. F. *The coral reef problem and isostasy* (Proc. Koninkl. Akad. v. Wetenschappen Amsterdam 19, 1916).
RITTMANN, A. *Vulkane und ihre Tätigkeit* (1936).
RITTMANN, A. *Die Entstehung des Sials und die Herkunft der Vulkanischen Energie* (Geologie en Mijnbouw, 1939).
RUTTEN, L. *On the age of the serpentines in Cuba* (Proc. Kon. Acad. v. Wetensch. Amsterdam, vol. 43, 1940).

STILLE, H. *Zur Frage der Herkunft der Magmen* (Abh. Preuss. Akad. d. Wiss. Jahrg. 1939, 1940).

UMBGROVE, J. H. F. *Verschillende typen van tertiaire geosynclinalen in den Indischen Archipel* (Leidsche Geol. Mededeel. 6, 1933).

UMBGROVE, J. H. F. *The relation between magmatic cycles and orogenic epochs* (Geol. Magazine, vol. 76, 1939).

VENING MEINESZ, F. A. *The mechanism of mountain formation in geosynclinal belts* (Proceed. Kon. Acad. v. Wetensch. Amsterdam, vol. 36, 1933).

VENING MEINESZ, F. A., UMBGROVE, J. H. F. and KUENEN, PH. H. *Gravity expeditions at sea* (vol. II, 1934).

VENING MEINESZ, F. A. *Evenwichtsverstoringen in de aarde.* (Inaugurale rede, Delft, 1939).

VENING MEINESZ, F. A. *Evenwichtsverstoringen in de aarde.* (De Ingenieur 1940).

VENING MEINESZ, F. A. *The Earth's crust deformation in the East Indies* (Proceed. Kon. Akad. v. Wetensch., Amsterdam XLIII, 1940).

VENING MEINESZ, F. A. *Gravity over the Hawaian archipelago and over the Madeira area; conclusions about the earth's crust* (Proceed. Nederl. Acad. v. Wetensch. 44, 1941).

VENING MEINESZ, F. A. *Gravity over the continental edges* (Proceed. Nederl. Acad. v. Wetensch. 44. 1941).

WATERSCHOOT VAN DER GRACHT, W. A. J. M. VAN, *Lateral movements on the Alpine Foreland of north western Europe* (Proceed. Kon. Acad. v. Wetensch. 41, 1938).

CHAPTER V

OSCILLATIONS OF THE SEA-LEVEL

"We have to account for periodic transgressional seas and orogenesis upon an enormous scale and clearly associated with vertical continental movements". (J. JOLY)

Introduction

The sea has repeatedly invaded large portions of the continents and then subsequently retreated. This part of the earth's history is an extremely monotonous one, but this very monotony, this continuous recurrence of the same events, constitutes one of its most important aspects. Why does the sea invade the continents, and why does the water later withdraw to the oceanic receptacles, only to advance again afterwards? In other words, in geological terminology, — what is the origin of the periodic trans- and regressions?

Before considering the major periodic movements, however, we should first note that there are undoubtedly examples of trans- and regressions of a mere limited, regional importance. In these cases it is but seldom possible to connect the relative shifts of the sea-level directly with one or more clearly discernible impulses.

In geosynclinal areas the phenomena may occur at a different rate as a result of their own and possibly opposed movements, and the negative or positive shifting of the sea-level may perhaps be quite counterbalanced. It is even conceivable that transgression may occur in a geosynclinal zone simultaneously with a negative phase, and vice versa. On the continents, too, the strand may advance or retreat over an area of greater or lesser extent as a result of different causes, but this need not necessarily indicate a world-wide movement.

A few such cases should be discussed before passing on to an examination of changes of world-wide importance.

Regional transgressions and regressions

The sea invaded the Baltic regions twice during the melting of the so-called Fennoscandian ice-sheet (fig. 48). The first to form as the ice-sheet retreated was the Baltic ice-lake, which occupied the southern part of the actual Baltic Sea (fig. 48 A). A strip of land in the south of Sweden became free as a result of the steadily increasing recession of the ice-front. This area was so low that it was flooded by the ocean, causing the Baltic fresh-water lake to be transformed into a sea — the Yoldia-Sea, which was

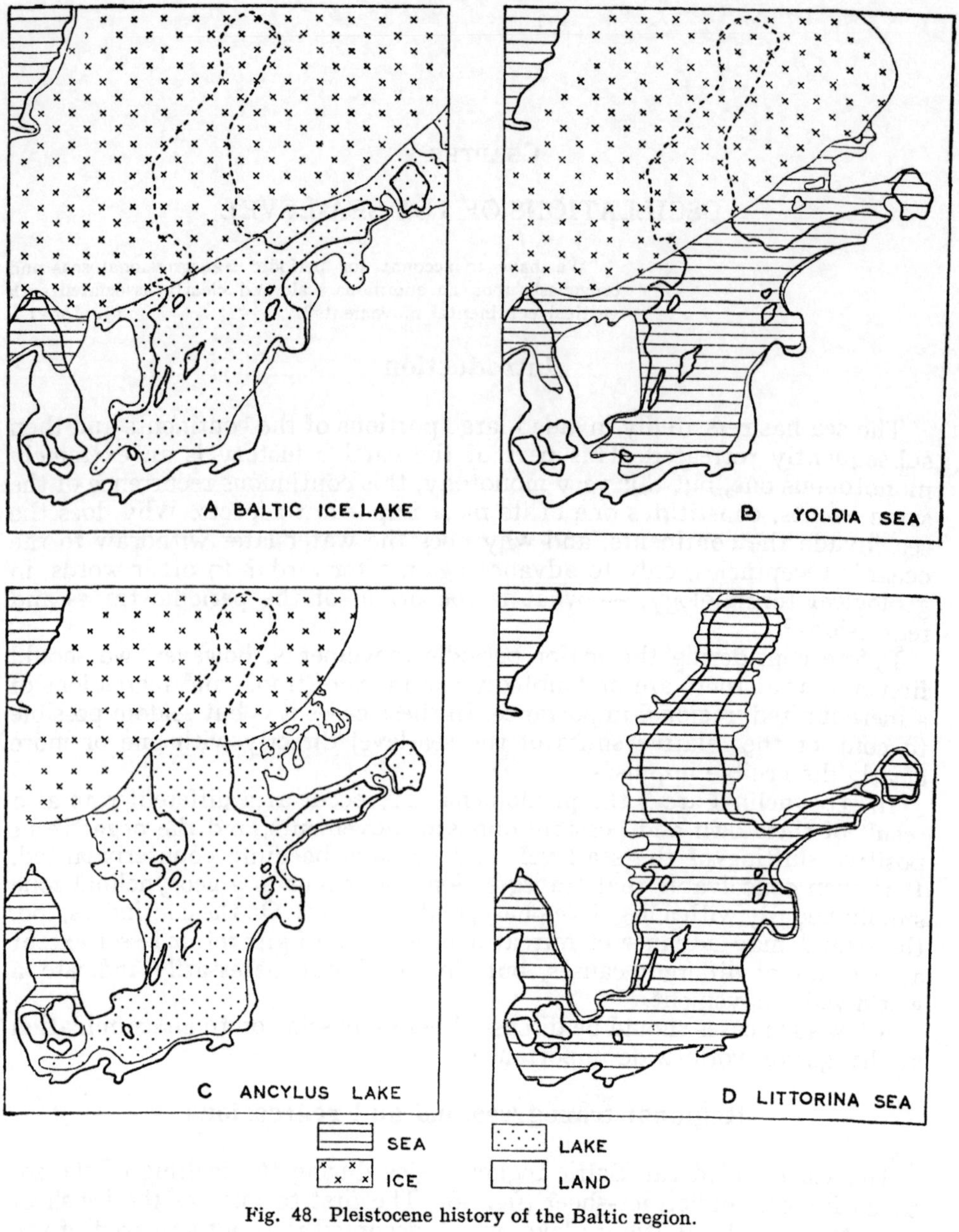

Fig. 48. Pleistocene history of the Baltic region.

connected with the Atlantic (fig. 48 B). The crust of South and Central-Sweden was now rid of the load of ice which had previously been pressing it down and began to rise gradually. The connection with the Atlantic was consequently severed, and the Yoldia-Sea was soon replaced by a fresh-water body — the Ancylus lake (fig. 48 C). While the crust had been strongly basined under the load of land-ice, the marginal belt around the glaciated tract had been super-elevated. As pointed out, the land began to rise slowly after the ice had melted. On the other hand, the peripheral belt — which includes the North of Germany and Denmark — began to subside. This, combined with the rising of sea-level by the melting of ice, caused the connection with the waters of the Atlantic to be re-established. The Straits of the Sund, the Great Belt and Little Belt were opened, and the fresh-water

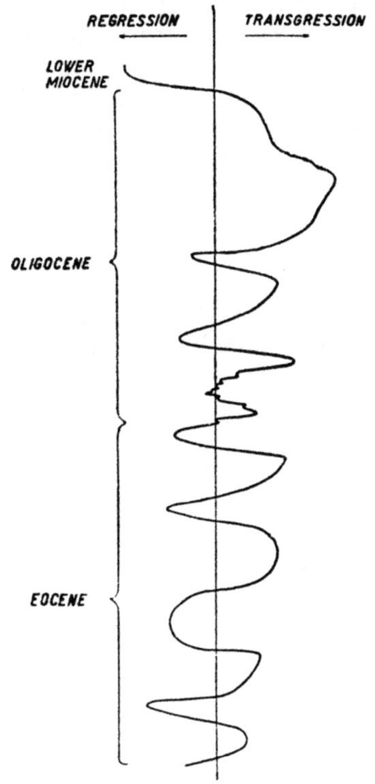

Fig. 49. Tertiary transgressions and regressions in the basin of Paris (After H. Stille).

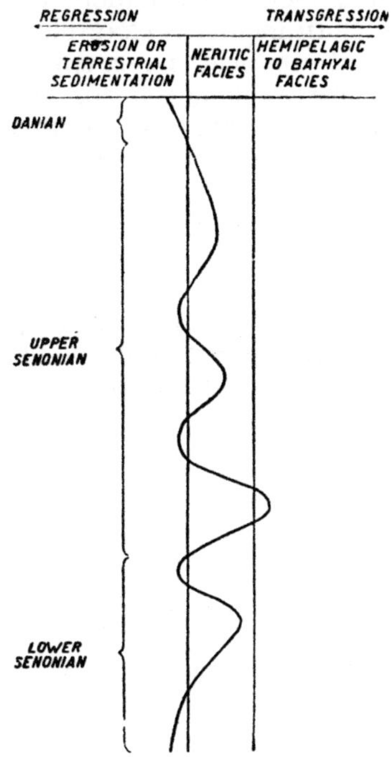

Fig. 50. Transgressions and regressions in the Upper-Cretaceous of Limburg, in the Netherlands.

of the Ancylus lake was replaced by salt water — the so-called Littorina-Sea (fig. 48 D), which was gradually transformed into the present Baltic-Sea.

I mentioned this example because it clearly illustrates how the local rhythm of the alternation of fresh and salt water can be derived from one constant, universal and non-periodic factor. The land movements are in fact a consequence of the retreat of the Fennoscandic inland-ice, while the rise of the sea-level is due to the melting of the Fennoscandic ice-cap and other ice masses. All this is brought about by one changing factor, viz. a world-wide increase in the average annual temperature.

This example clearly shows that the repeated advance and retreat of the sea in a restricted area need not necessarily be the reflection of a universally active cause that changes periodically. The curve in fig. 49 indicates the successive trans- and regressions which affected the basin of Paris. If we bear in mind that the basin has a complicated history (see Chapter III), it will become clear that the latter has also found expression in the curve. In many cases the origin of trans- and regressions remain a matter for conjecture. To what cause or causes must we for instance attribute the oscillations of the sea-level in South-Limburg during the Senonian (fig. 50)? We have to content ourselves with more or less vague surmises. In a very readable article, Born mentions examples of great differences in amplitude and time of such rhythmical movements. It seems as yet impossible to furnish a satisfactory explanation for all these phenomena. When considering a world-wide "rhythm", we should not forget that numerous local movements, which may have had a very different origin, have undoubtedly also played a part in several regions.

World-wide transgressions and regressions

Large transgressions, however, are undoubtedly known to have occurred at a time when there could be no question of a considerable change of sea-level under the influence of the melting and extension of vast masses of ice,

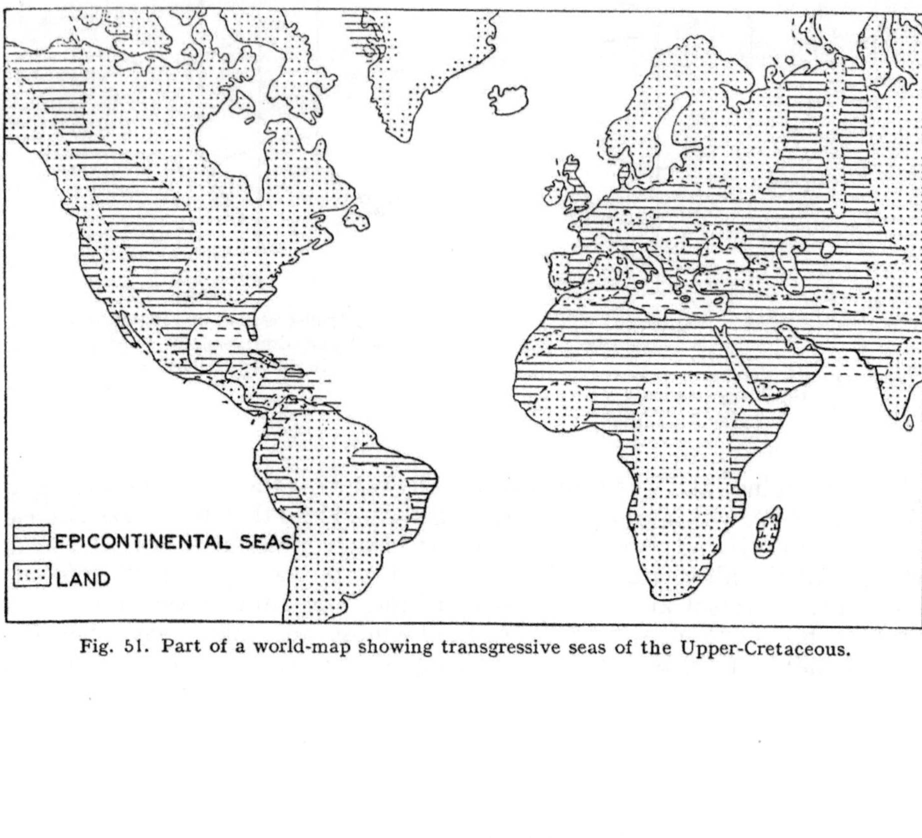

Fig. 51. Part of a world-map showing transgressive seas of the Upper-Cretaceous.

the considerable expanse of which only proves that they cannot merely be explained by the movement of parts of a continent.

An example of this is furnished by the epicontinental seas of the Upper-Cretaceous (fig. 51). The gradual advance of the sea into the European Continent can be closely followed. This great invasion of the sea was not only felt in Europe, but also on other continents. Conversely, a world-wide retreat of the sea prevails towards the close of the Mesozoic, and is in turn succeeded by a positive phase of the epicontinental seas in the Lower-Tertiary. Suess was the first to show that these phenomena have a world-wide importance and that the periods of transgression were far longer than the relatively short periods of regression.

Suess, Haug, Stille and Grabau paid special attention to the chronological and regional distribution of the transgressions, and each in turn came to the conclusion that a great number of major transgressions took place each being separated by periods of widespread emersion of the continents. There was a rhythmic advance and retreat of the sea. We can therefore only conclude that the trans- and regressions on the continents must be ascribed solely to a world-embracing cause. Stille expressed the synchronism of the great trans- and regressions in his law of epirogenic synchronism, which Bucher formulated as follows: "In a large way the major movements of the strandline, positive and negative, have affected all continents in the same sense at the same time".

Once this rule is established statistically, the local and regional exceptions appear clearly. Thus Stille gives two curves for the Triassic, one representing the world-wide movement of the sea-level, and the second the strongly diverging and even opposed transgressions and regressions of the Triassic in Germany (fig. 52).

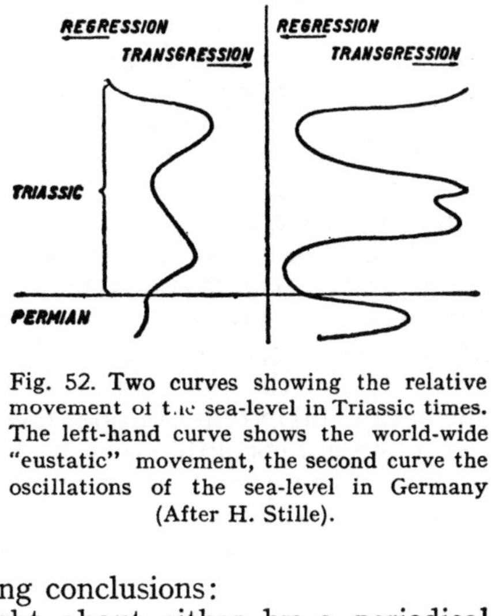

Fig. 52. Two curves showing the relative movement of the sea-level in Triassic times. The left-hand curve shows the world-wide "eustatic" movement, the second curve the oscillations of the sea-level in Germany (After H. Stille).

I discussed the problem of trans- and regressions in a recent paper and will therefore deal with them very briefly at present. A study of the oscillating sea-level caused me to draw the following conclusions:

(1) World-wide regressions are brought about either by a periodical elevation of the continents, or by periodical subsidence of the ocean-floors, or by simultaneous but opposed movements of both continents and ocean-floors. The last possibility appears to be the most probable one, and the same applies — *mutatis mutandis* — to the transgressions.

(2) Epochs of folding coincide with periods of world-wide regression, which were relatively short in comparison to the much longer intervening periods of transgression.

(3) Movements of the sea-level and folding are caused by subcrustal processes, which elapsed periodically, as it were, with a rhythmic cadence.

(4) Other phenomena (the magmatic cycles, the formation of basins and dome-shaped elevations, and the alternating decrease and increase in compression) are closely related to this "rhythm", and they, too, show that their origin must be sought deep within the earth, as pointed out in the preceding chapter.

In Chapter IV, I explained that my conception of the structure of the earth's crust coincides with that of Daly and Vening Meinesz. This means that the earth is actually enveloped by a rigid crust. A cross-section of a continent shows that the crust is composed of sial and crystalline sima, and that a thinner sialic layer (though resting on a thicker layer of sima) is present under the Atlantic, and that a very thick layer of crystalline sima extends under the Pacific. Wegener's view that sima represents a viscous substratum in which continents would only drift as rigid blocks is unacceptable under the present day circumstances [1]). It may still be conjectured whether the rigid and crystalline upper layer of sima enveloping the earth at present had at some time been considerably thinner or possibly non-existent (aside from the fact as to whether they were controlled by periodic radio-active processes, as Joly assumed, or whether another unknown internal process was responsible for them). At any rate, the periodicity of trans- and regressions shows that the differences in level between continents and ocean-floors have changed periodically!

The curve in Table II indicates the result as regards the relative movement of the level of the sea, showing the positive and negative shifting of the sea-level and its chronological relation to epochs of compression. This curve is entirely schematic. Besides, it is highly probable that certain trans- and regressions, which cannot be indicated for lack of data, had a greater amplitude than others.

A detailed study of the facies of the strata deposited on the continents by transgressive seas may produce evidence of the height of the sea-level at various periods above certain parts of the continents. But in that case local movements (e.g. of basin-shaped depressions) will also have to be taken into account, and the problem is thus far from being a simple one. The vast epicontinental extent of upper-cretaceous strata, which are often compared to the present cocolith-ooze found at a depth of at least 1,000 meters, would lead one to suppose that an important shift of the strandline accompanied some transgressions. Kuenen assumed that a major transgression would result from an eustatic movement with an amplitude of some 40 meters. This amount, as he himself declared, should be regarded as a minimum value. We might add that it ought undoubtedly to be more. For Daly pointed out that if all the ice of the ice-caps were to melt, this fact alone would already cause the sea-level to rise some 40 to 50 meters [2]). We may safely say that the period in which the last world-wide regression reached its deepest level now lies behind us, but we are undoubtedly still far removed from the day when the prevailing transgression will have reached its maximum. During the maximum glaciation in the Pleistocene, the sea-level was approximately 100 meters lower than it is now, and the sea was already

---

1) See Chapter VI, p. 96 and 107. 2) See Chapter VII, p. 118.

in a new transgressional stage at that moment. Thus the maximum for a transgression must be fixed at at least 150 meters, plus an unknown amount, and the rise of the sea-level towards the maximum of the now prevailing transgressional phase will attain 50 meters above the present level, plus an unknown but undoubtedly considerable amount. Penck estimated that transgressions were caused by a relative rise of the sea-level with an amplitude of 500 meters, and Joly evaluated the amount of the eustatic change between a deep regression and an extreme transgression at 1,200 meters.

Many other factors complicate this problem. For instance, the formation of the Mediterranean and the East and West-Indian and Asiatic basins would have resulted in an eustatic lowering of the sea-level of approximately 60 to 80 meters according to Kuenen. The most important question concerns the depth to which the sea-level was depressed in distinct periods of intensive regression, in other words the extent of the change to which the distance between the surface of the continents and the ocean-floors was subjected during the pulsating rhythm of subcrustal processes. Joly was the only one who approached this question from the geophysical side and he arrived at an order of 1,000 meters. No one can foresee what further geophysical speculations may lead to eventually, but it is clear that this question has an extremely important bearing on various geological problems. For if the sea-level moved downwards far beyond its present position during certain periods, it would have had a strong influence on many phenomena. In this connection may be mentioned the origin of isthmian links, and eventually broader transoceanic landbridges. Further, all those physiographic factors that lead to glaciation would be considerably accentuated. It will be evident, moreover, that new aspects would be opened as regards the problems of submarine canyons and the origin of barrier reefs and atolls.

Obviously, not a single marine sediment could have been deposited on the continents or within shoal-water geosynclines during such a large emersion of the continents as may be expected to have preceded the late paleozoic and the pleistocene glaciation. As far as I have been able to ascertain, however, the actual facts do not confirm the idea of a world-wide regression of such extent that the surface of the sea would at some time have been lowered to a few thousand meters below its present level, but this important question deserves constant attention.

References

BARRELL, J. *The strength of the Earth's Crust* (Journal of Geology 22, 1914).

BARRELL, J. *Rhythms and the measurement of geologic time* (Bull. Geol. Soc. of America 1917).

BORN, A. *Periodizität epirogener Krustenbewegungen* (Rep. 16th Intern. Geol. Congress Washington 1933, 1, 1936).

BUCHER, W. H. *The deformation of the Earth's Crust* (1933).

DALY, R. A. *The changing world of the ice age* (1934).

GRABAU, A. W. *Oscillation or pulsation* (Rep. 16th Intern. Geol. Congress Washington 1933, 1, 1936).

GRABAU, A. W. *Palaeozoic formations in the light of the pulsation theory* (1936).

HAUG, E. *Les géosynclinaux et les aires continentales; contribution a l'étude des transgressions et regressions* (Bull. Soc. Géol. de France, 3e ser. 18, 1900).

JOLY, J. *The surface history of the Earth* (Oxford Univ. Press. 1925).

KUENEN, PH. H. *Quantitative estimations relating to eustatic movements* (Geologie en Mijnbouw 1, 1939).

KUENEN, PH. H. *Causes of eustatic movements* (Proceed. sixth Pac. Sci. Congress II, 1940).

PENCK, A. *Theorie der Bewegung der Strandlinie* (Sitzungsber. Preuss. Akad. Wiss. 1934).

SCHUCHERT, CH. *Correlation and chronology on the basis of paleogeography* (Bull. Geol. Soc. of America 26, 1916).

STILLE, H. *Studien über Meeres- und Bodenschwankungen* (Nachr. K. Ges. Wiss. Göttingen. Mat. -phys. Kl. 1922).

STILLE, H. *Grundfragen der Vergleichenden Tektonik* (1924).

SUESS, ED. *The face of the Earth* (*transl. Sollas*) 2, 1906.

UMBGROVE, J. H. F. *The amount of the maximal lowering of the sea-level in the pleistocene* (Proc. Fourth Pacific Science Congress, vol. 2, 1930).

UMBGROVE, J. H. F. *On rhythms in the history of the Earth* (Geol. Magazine 76, 1939).

Chapter VI

THE FLOOR OF THE OCEANS

"....it appears that the higher continental masses and the depressed oceanic basins came into being very early in the history of the earth"

(Ch. Schuchert)

Introduction

The antipodal distribution of continental blocks and oceanic receptacles ranges among the most peculiar features of the earth. The deep-sea area of the North Pole is situated directly opposite the Antarctic Continent. The six non-polar continents are grouped in pairs, with a roughly meridional orientation, and form an elongated triangle with its apex pointing southwards. Conversely, the oceanic sectors extend in an opposite direction. The broad bases of the continents, arranged in a nearly continuous ring around the North Polar Basin, have their antitype in the broad, circum-Antarctic ring of water. The counterpart of the Indian Ocean is found in North-America and that of Australia in the Atlantic north of the equator, etc. Of course, there are exceptions to this antipodal distribution of land and water, but the fact remains that only 1/20 of the whole land-surface is antipodal to land [1]). This exceptionally prominent feature of the globe, in which land masses are opposed to oceanic areas, reminds one of a tetrahedron, i.e. a crystallographic body in which a rib is always situated opposite a plane. It has given rise to a great deal of speculation, but it need hardly be said that this characteristic is such a remarkable one that any valid theory as regards the origin of continents and oceanic receptacles ought to be able to furnish an explanation of their tetrahedral arrangement [2]). It is clear, however, that this question involves the whole intricate problem of the origin of continents and ocean-floors, and there are many conflicting views on this subject. Some are of the opinion that the continents and oceans represent permanent features. Another hypothesis, diametrically

---

1) Greenland and the Arctic Islands of North-America are situated opposite Victoria and Wilkesland. New-Zealand is antipodal to the Iberian Peninsula, Patagonia to part of North China and Grahamland to the Taimyr Peninsula.

2) The geometric pattern of the surface of the earth gave rise to the well-known tetrahedral theory of Lowthian Green. Among the later exponents of similar theories should also be mentioned Gregory, de Lapparent, Arldt and Kober. For a discussion of these hypotheses we refer the reader to "The unstable Earth", by J. A. Steers (Methuen, London 1932) and to Bucher's "The deformation of the earth's crust" (1933).

opposed to the preceding one, is that the oceanic receptacles originated as a result of the submergence of land-masses. Others associate their formation with a drifting-apart of continents, while a fourth group claims that at least two of the ocean-floors were formed by stretching of continental blocks. All these possibilities will be discussed below, but it appears advisable to begin with a brief review of all that is known at present of the bottom relief of the oceans, while a comparison with the results of geological and geophysical research is obviously warranted at this stage.

The major characteristics of the bottom-relief

As more and more data become available, it becomes increasingly evident [1]) that the relief of the ocean-floors presents some very curious characteristics.

A comparison between the various oceanic sectors will make it immediately clear that the relief of the Atlantic and Indian Oceans —while differing to a considerable degree from that of the Pacific, especially in the case of the North Pacific basin proper, i.e. the area limited by the andesite line and a southern demarcation running approximately from the Fiji to the Galapagos-Islands — concur in many respects. The morphology of these areas is matched by their geological structure, for seismic and petrographic data both clearly show that the bottoms of the Atlantic and Indian Oceans are very different from that of the Pacific. The generally accepted view at present is that no sialic layer exists beneath this Pacific sector, though such a layer is in fact assumed to extend under the Atlantic and the Indian Ocean.

These broad outlines make it possible to divide the morphological characteristics into four major groups:

(1) The Atlantic Ocean, and the western part of the Indian Ocean, with their numerous and comparatively flat-bottomed basins, separated by relatively narrow and steep ridges.

(2) The large North Pacific basin, with its frequent linear and at times intersecting ridges and throughs. No mention will be made of the manifold basins in Eastern and Southeastern Asia, since these formations are situated inside the andesite line. Nor will we discuss Melanesia, which comprises the East-Indies from a geological and morphological point of view. Most authors assert that the andesite line represents the true boundary of the Pacific basin proper, and the real problem may thus be stated to begin in this area. Thus too, when dealing with the problem of the Atlantic, no mention will be made of the Mediterranean and West Indian basins. Moreover, these regions were already discussed in Chapter III.

(3) The eastern part of the Indian Ocean, the south-western part of the Pacific and the three Antarctic basins.

---

[1]) The following considerations are based on the deep-sea chart of the Atlantic by Stocks and Wüst (1935), Schott's chart of the Indian Ocean (1935), the chart of the John Murray expedition for more recent data on this area, a chart by Leahy (1938) and another fine chart of the Pacific by the U.S. Hydrographic Office (1939). A clear picture of diverse morphological questions will furthermore be found in publications by such authors as Cloos, Wüst, Schott and Mecking.

These basins are in so far similar that a plain bottom-relief with hardly any subdivision is found in these areas. The three elongated deep-sea basins around the Antarctic continent (fig. 53) may be referred to briefly as the Atlantic, the Indian and the Pacific Antarctic basin [1]).

The above conception of the basins' morphology may possibly prove

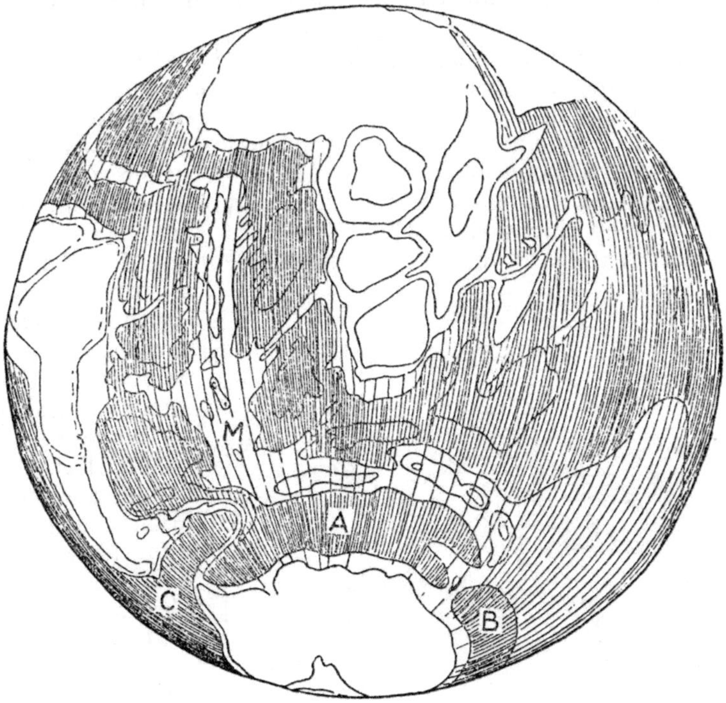

Fig. 53. Sketch of the submarine relief of the Atlantic and Indian Ocean floors surrounding Antarctica (After H. Cloos) A, Atlantic Antarctic Basin; B, Indian Antarctic Basin; C, Pacific Antarctic Basin; M, Mid-Atlantic Rise.

to be more complicated than we suppose it to be, as neither basin has been examined in detail.

The eastern part of the Indian Ocean and the south-western part of the Pacific form vast receptacles with a symmetrical arrangement in respect to Australia and Melanesia, the concave sides being turned towards one another.

(4) The North-Polar basin. Little is known of the details of this receptacle (cf. the basins of group (3), which it would appear to resemble in some

[1]) The Atlantic Antarctic basin extends eastwards from Grahamland to the southern point of Africa. It is separated from the Indian Antarctic basin by the Kerguelen Rise, and reaches as far as the ridge connecting Tasmania with Antarctica. It is succeeded by the Pacific Antarctic basin, which — extending eastwards to South-America and Grahamland — is bounded on its northern extremity by the long, so-called East-Pacific Rise.

respects), and no further reference will consequently be made to this formation [1]).

On the other hand special attention will be paid [2]) to the oceanic sectors of groups (1) and (2).

The Atlantic Ocean and the western part of the Indian Ocean

We will begin with a summary of a series of features which illustrate the significant congruence of these oceanic sectors and the surrounding continents, showing that in many respects the topographic limits between continental and oceanic areas do not correspond with structural limits.

The regions appearing under group (1) are characterized by a great number of deep-sea basins with comparatively flat bottoms surrounded by relatively steep ridges. The whole structure is in many respects very similar to that of the intervening African continent (fig. 54). This similarity is further accentuated by the fact that Africa, besides containing many more basins than South-America, is also flanked by far more submarine basins than the American sector west of the Mid-Atlantic Rise. The morphology of the Southern Atlantic blends, as it were, with that of the Atlantic Antarctic basin (fig. 53 and 55). The Mid-Atlantic Rise bends eastwards in the same way as the loop of the Southern Antilles and ultimately forms the boundary of the Antarctic basin, joining the Kerguelen Rise. In the Indian Ocean an analogous ridge bends eastwards around the Indian Antarctic basin.

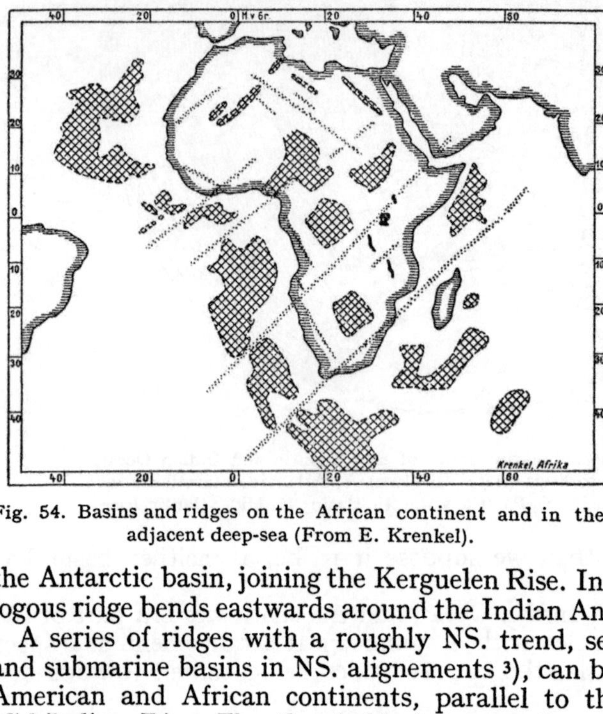

Fig. 54. Basins and ridges on the African continent and in the adjacent deep-sea (From E. Krenkel).

A series of ridges with a roughly NS. trend, separating the continental and submarine basins in NS. alignements [3]), can be observed on the South-American and African continents, parallel to the Mid-Atlantic and the Mid-Indian Rise. The thresholds between the continental basins consist mostly of precambrian rocks. Two main strikes dominate in Africa since the

[1]) This area's latest bathymetrical chart can be found in a publication by Stocks (1939). An earlier map by Holtedahl shows the geology of the surrounding continents.

[2]) I purposely avoided submarine canyons when dealing with the problem of the ocean-floors. As long as the origin of these phenomena (to which increasing attention has been paid of late) remains unsolved, they should be kept out of any discussion of the problem of the origin of ocean-floors.

[3]) Krenkel's configuration has been followed.

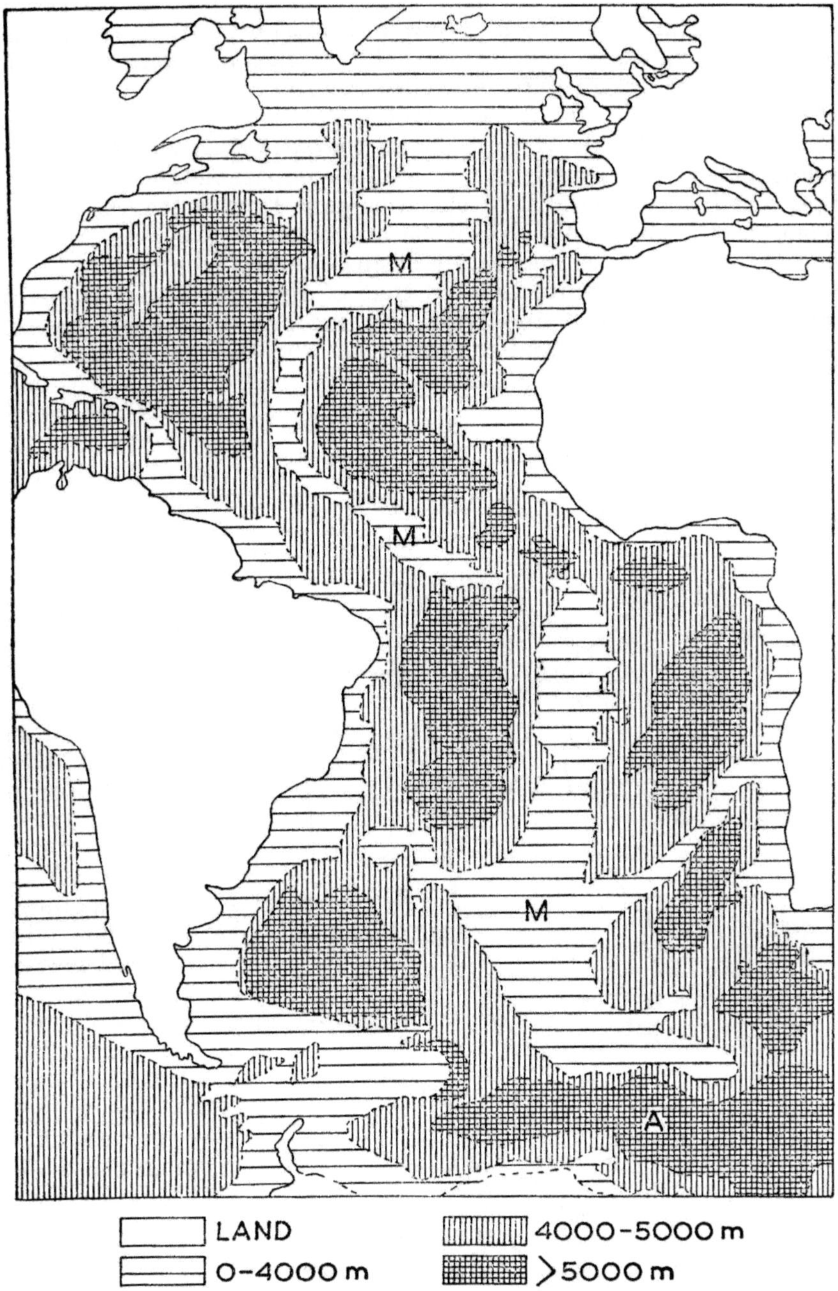

Fig. 55. Bathymetric chart of the Atlantic Ocean. M, Mid-Atlantic Rise; A, Atlantic Antarctic basin.

Cambrian: a NE.–SW., or so-called Somali strike, and a NW.–SE., or so-called Erythraean strike. Both played an important part in the whole later history of the continent. The same strike is again apparent in the narrow submarine ridges separating the deep-sea basins (fig. 55). The same holds good for the northern part of the Atlantic. Bucher [1]), for instance, observes: "The map of the North-Atlantic shows that the rises which separate the individual basins consist of lines of swells comparable to a certain extent to that which in Eastern North-America runs from southwestern Ontario through the Cincinnati and Nashville domes, separating the Appalachian from the East-central sedimentation basin. The whole pattern of the ocean-floor seems, in fact, comparable to that of the basins and swells of the continental areas outside the great orogenic belts, although the scale is larger both horizontally and vertically on the oceanic surfaces".

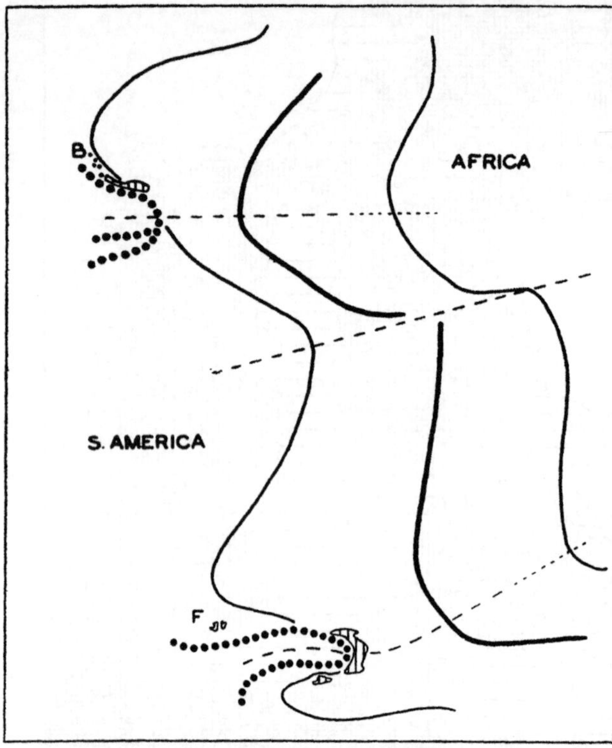

Fig. 56. Symmetry of the Northern and Southern Atlantic. (After H. Stille).

Mention should be made here of the opinion of Cloos[2]), who asserts that a domeshaped elevation with a fault-system, accompanied by volcanism and similar to that observed on the European and African continents, exists in the Azores, and this leads him to conclude that there can be no fundamental difference between this part of the Atlantic and a continent.

Another striking feature, as pointed out by Stille (who illustrated it most suggestively in one of his recent papers) is the symmetry of the northern and southern Atlantic sectors (fig. 56). These may be described as forming one another's counterpart. This applies to the submarine characteristics (e.g. the convex arc of the Mid-Atlantic Rise bends eastwards twice), as well as to the features above sea-level (e.g. the arc of the

[1]) Bucher, 1940, p. 492.

[2]) Cloos, Geol. Rundschau, 30, 4a, 1939.

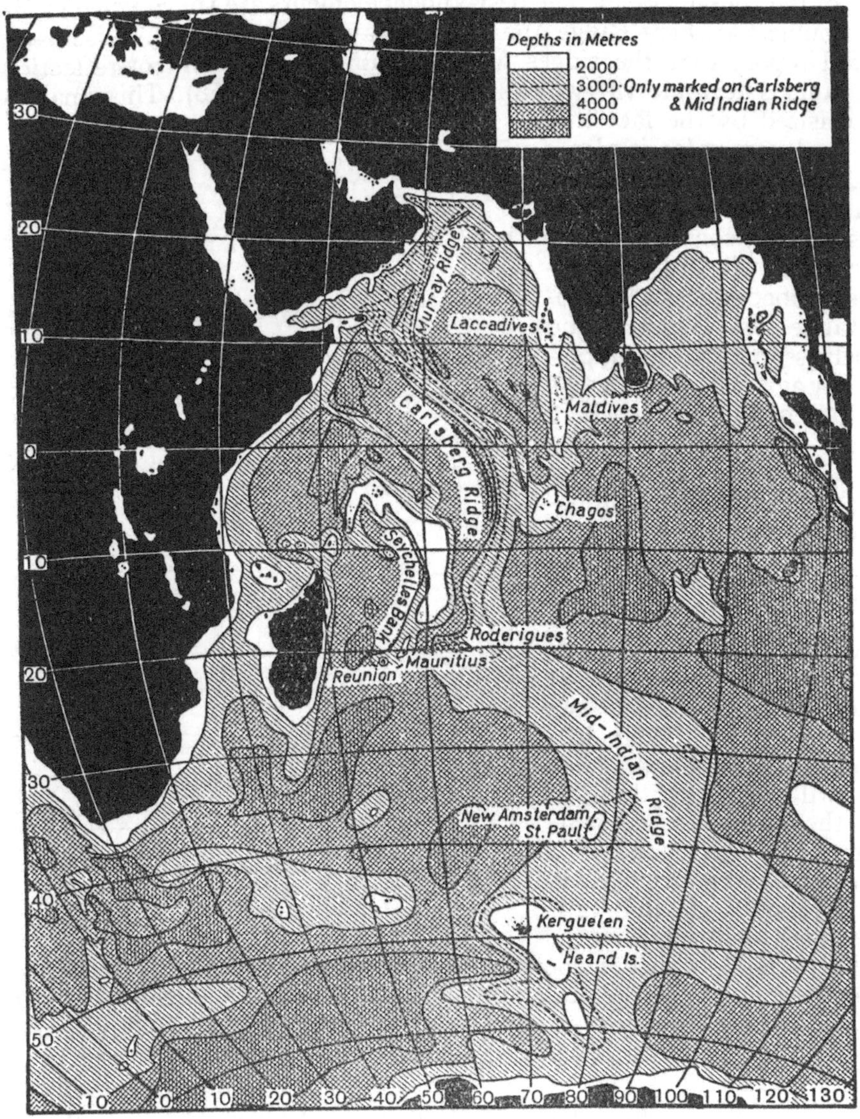

Fig. 57. Bathymetric chart of the Indian Ocean (from J. H. Wiseman and R. B. Seymour Sewell).

northern Antilles with the shelf of the Bahamas and the arc of the southern Antilles with the Falkland-Islands shelf, etc.).

Symmetrical features are also found in the Indian Ocean, and deserve special attention, as half of the symmetrical structure appears on the African continent. The new chart of the John Murray expedition shows the existence of a series of widely-curving ridges in the north-western part of the Indian Ocean. The best-known elements are the Seychelles Bank, the double Carlsberg Ridge and the Murray Ridge (fig. 57). Wiseman and Sewell report that the ridges form the almost exact mirror-reflection of the famous tectonic rift-valleys in East Africa (fig. 58). This analogy is emphasized by the fact that the Carlsberg ridge is also a belt of tectonic earthquakes and volcanism (fig. 59). This consequently constitutes another point of similarity between suboceanic and continental sectors. Wiseman and Sewell assume that an upper-tertiary age may probably be attributed to the Carlsberg Ridge. The Murray Ridge probably fuses with the Carlsberg System, and this probably connects in the south with the ridges upon which the Lacadive-, Maldive- and Chagos-Islands rest. The latter appear to unite as the Mid-Indian Rise, which can be traced to Antarctica via New-Amsterdam, the Kerguelen, and the Heard-Islands.

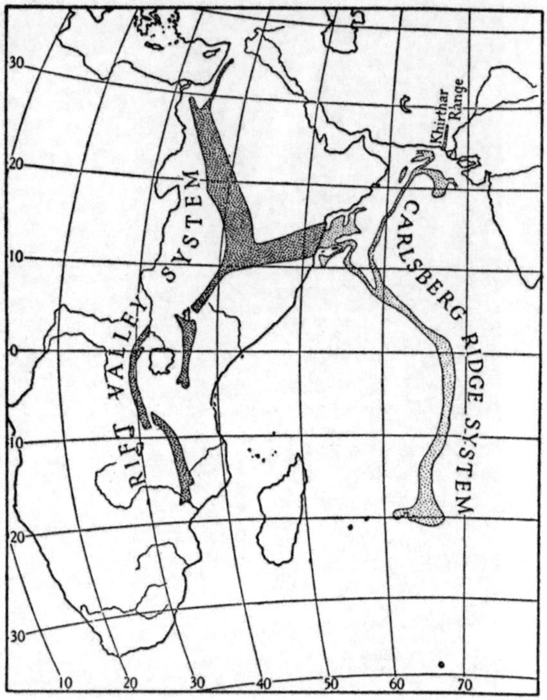

Fig. 58. The systems of the Carlsberg ridges and the Great Rift-Valley. (From J. H. Wiseman and R. B. Seymour Sewell).

It will be clear that, in case the basins of the north-western area correspond with old structural elements, their frames should at any rate have been subjected to a "rejuvenation" (a phenomenon also observed on the continents) in comparatively recent times. The same applies *mutatis mutandis* to the relief of the Atlantic. Young fault-systems, graben, upper-tertiary strata and pleistocene volcanism are found in Iceland, the Azores and St. Paul. Unfolded upper-tertiary sediments occur in the first two areas. Other young volcanic regions are Ascension, Tristan da Cunha, St. Helena, Gough-Island and Bouvet-Island. Moreover, the seismic unrest of the Mid-Atlantic Rise proves that the movements are still being continued [1]).

Petrographical data obtained in the oceanic areas stress the results of geological and seismic observations and point to the sialic composition of the floor [2]).

---

1) cf. Bucher, 1940, p. 494–495.

2) Plutonic rocks are known to occur in some small islands of the Indian Ocean, e.g. in Juan de Nova (between Africa and Madagascar), the Seychelles, Socotra and the western sector of the Kuria Muria-Islands (granite, syenite), while dubious finds (which were not actually discovered as bedrock) have been reported to

So far attention has chiefly been paid to the pattern of the submarine relief. Yet it is obvious that valuable information can also be obtained from the geology of the surrounding continents. We will consequently give a brief summary [1]) of a few of the most important points that call for special attention when dealing with the problem of the ocean-floors.

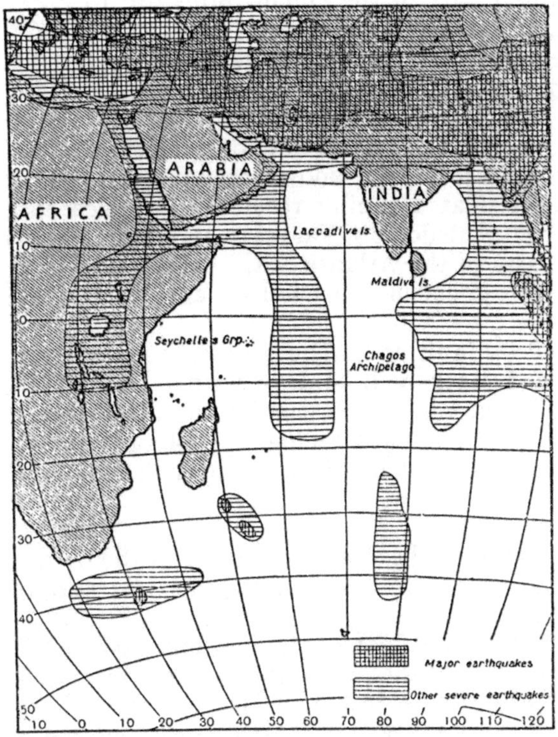

Fig. 59. The earthquake belts of the Indian Ocean and its surroundings (After Heck).

(*a*) Scandia. This area lay above sea-level till the Lower-Tertiary according to Holtedahl and Frebold. Its former extent is purely hypothetical.

(*b*) Appalachia extended at least 200 miles beyond the eastern coast of North-America (cf. Barrell, Schuchert and Ewing), and now lies submerged to a depth of at least 3,000 to 4,000 meters. Geophysical data seem to imply that this area began to founder in the Triassic and that this process continued until at any rate the Tertiary (fig. 22 and 23).

(*c*) The seismic method of investigation was also applied by Bullard and Gaskell to the submarine geology of the eastern part of the Atlantic i.e. along a line extending 170 miles WSW. from Cape Lizard. In a preliminary paper these authors revealed the presence, in this area, of a submerged block probably consisting of igneous rocks covered by triassic strata.

(*d*) In his monograph on the Congo basin Veatch writes as follows: "A consideration of the source of the sediments forming the Lubilash of the western Congo, likewise, points to a more elevated, though not necessarily

be present in the Komores. Clay and chlorite schists have been observed in Mauritius, and dolomitic limestone, tertiary petrified wood, mica-diorite, felsite-porphyry, labradorporphyrite, schists and siliceous slate in the Kerguelen. Miocene and eocene limestone rest on basalt and trachyte in Christmas-Island. Only a few indications of these rocks are found in the Atlantic. Granitic xenoliths occur in the volcanic lavas of Ascension and probably in Tristan da Cunha. Plutonic rocks, sediments of varying age and metamorphic strata are found in a few of the Cape Verde-Islands and the Canaries.

[1]) The geology of the coastal areas was discussed in Chapter II, and is also dealt with in the Appendix.

mountainous, land mass occupying the present depressed coastal plain region and extending into what is now the Atlantic Ocean."

(*e*) Faulting of presumably pliocene or post-pliocene age occured along the coast of Africa, Arabia and India around the Arabian Sea, and movements of a similar age caused part of the Oman-Indus arc to break down into the Arabian Sea according to Wiseman and Sewell.

(*f*) The southern part of the South-African Variscides and the probable connection with the Cedar Mountains disappeared into the sea. Nothing is known of the original extent of this land mass.

In none of these cases need one think of blocks of enormous continental extent, though many investigators are indeed known to have jumped at this conclusion, while others arrived at a similar result on the plea that the folded chains (the Precambrian, Caledonian, Variscian and Alpine belts) terminate abruptly along the coasts of the Atlantic. Still others base their conclusions on the analogy of the submarine morphology and that of the continents [1]).

Many coasts are indeed formed by faulting [2]). The possible consequences of the idea of tremendous subsided blocks will be dealt with subsequently. Biogeographical data will play no part in this discussion, as these should never be allowed to influence the solution of any geological problem *a priori*, even though such data may prove to be valuable if compared with geological results.

Recapitulating, a comparison between the submarine relief of the oceans and the geology of the surrounding continents may therefore be said to lead to the following conclusions:

1. The bottom of the Atlantic and the floor of the western part of the Indian Ocean are built up of similar (sialic) material as the continents.
2. The same events (i.e. the formation of basins, ridges, and eventually rift-valleys) occured in the oceanic sectors as well as the bordering continents.
3. The structural lines determining the location of the continental and submarine basins date from very early precambrian times.
4. The submarine relief, however, indicates that it underwent the same process of "rejuvenation" in comparatively recent times as the continents.
5. The submarine basins need not have originated simultaneously, any more than the continental basins, which are known to have formed in very different periods.
6. The pattern resulting from the arrangement of the three oblong basins around Antarctica, the blending of the submarine relief of the adjoining areas and the symmetry of the bordering oceanic sectors must surely have some special significance. Cloos attributed this constellation to the influence of the rotation of the earth. It would in any case appear to reflect the activity of

[1]) Stille, for instance, speaks of a 'Destruktionsfeld, ein Feld der Einsenkung und des Einbruchs, herausgeschnitten aus dem uralten Südkontinent der in Resten heute noch östlich und westlich von ihm aufragt".

[2]) Born writes: "Es ist das Lissaboner becken nicht anders als eine Teilerscheinung der in Abbröckeln begriffenen West Europa". (Born 1932, p. 698).

subcrustal forces, which would then have produced this pattern in the early Precambrian.

7. The arrangement of the submarine basins and their increasing depth and width towards the south until confronting the Atlantic and Indian Antarctic basins on a broad line, reveal two large wedge-shaped oceanic sectors with their apexes pointing northwards, while the South-American, African and Australian continents point in the opposite direction. There is no doubt in my mind that this situation can be said to constitute a fundamental feature, resulting from primordial and deep-rooted forces. The same applies to the tetrahedral arrangement of the continents and their antipodal relations to the oceanic depressions as mentioned in the introduction.

8. Several blocks of unknown size have foundered very deeply. Many coasts are bounded by faults. Nevertheless, the characteristics mentioned under 1–7 should not induce us to suppose that the whole floor of either the Atlantic or the Indian Oceans constitute foundered continental blocks. However, this possibility will be examined below.

9. The oceanic and continental sectors differ chiefly as regards the thickness of the sialic layer. This layer is thinner under the first sectors, and situated at a great depth. It can be stated to constitute one of the major problems of our globe.

The Pacific Ocean

The large North-Pacific basin, situated north of an imaginary line running from the Fijis to the Galapagos-Islands along the northern margin of the Albatross Plateau, is characterized by many linear ridges and troughs which occasionally intersect one another, thus giving rise to a pattern which is met with in no other oceanic area. This linear arrangement, which is such a prominent feature of many Polynesian islands (fig. 60), is believed to have been caused by the rising of magma along faults and fissures in the floor of the Ocean, which in this case appears to be composed of basic rocks (crystalline sima).

Many volcanic peaks in the central part are crowned with atolls, and nothing is consequently known of the nature of the eruptive rocks in their foundation. A quantity of data could nevertheless be obtained outside this area, upon which many investigators — especially Chubb and Lacroix — based conclusions of a general kind. Chubb distinguishes between nepheline-bearing and nepheline-free areas, and assumes that the atolls, too, were formed on top of nepheline-bearing volcanoes. The following quotation will illustrate the fundamental thought underlying Chubb's line of reasoning and conclusions (fig. 61): "The lavas of the nepheline-free zone closely resemble those of the nepheline-bearing zone, except that they contain enough silica to saturate the alkalies. Many indeed contain a fair proportion of virtual free silica. Perhaps a similar magma fed them all, but in the nepheline-free zone it has to ascend through a certain thickness of highly siliceous rocks. If so it might be expected that some of the latter would be absorbed, and would produce exactly these differences in the lavas that are actually found.

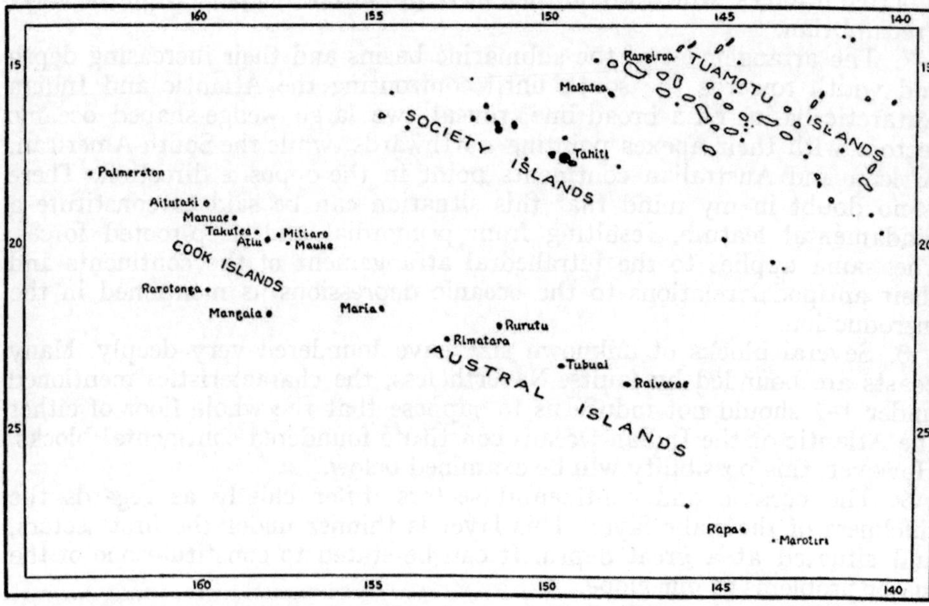

Fig. 60. Linear arrangement of island groups in the Pacific (From L. J. Chubb).

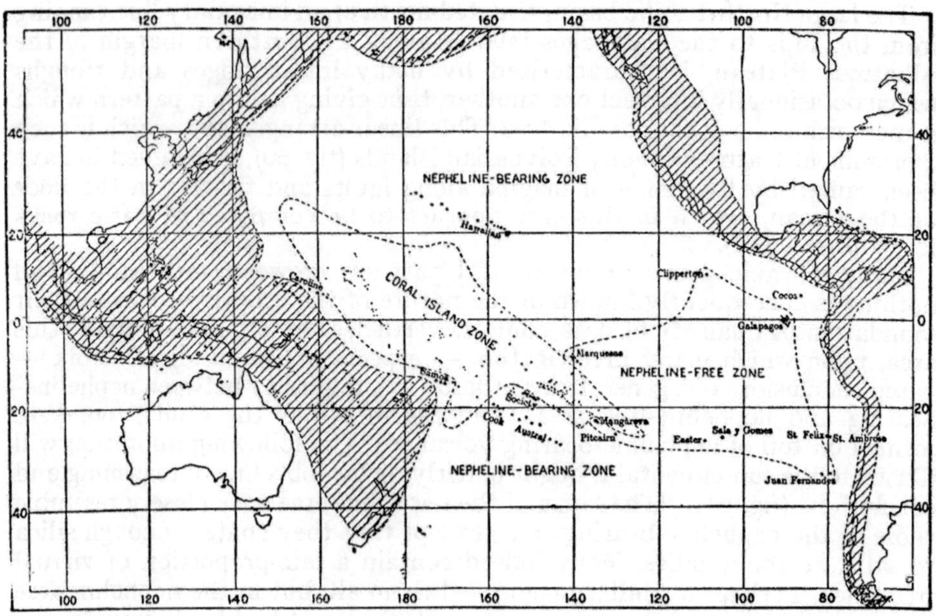

Fig. 61. The Pacific Ocean, showing distribution of rock types. Oblique shading: andesitic zone. (From L. J. Chubb).

For this reason it is suggested that the Albatross Plateau is a submerged block of sial".

The last suggestion has a direct bearing on the problem of areas which have eventually been submerged in the Pacific. Born [1]) had previously already observed: ... "von dem Einbruch alter Massen kann hier nicht die Rede sein. Die einzige Ausnahme könnten Albatrossplateau und Osterinsel, also Amerika benachtbarte Gebiete bilden". In a recent paper on this island, Bandy observes that "continental masses had no part in the formation of Easter-Island, and the island stands as an argument against continental masses existing in this part of the Pacific. There still remains, however, the question of the Albatross Plateau upon which Easter-island rests." It can not yet be said for certain whether a thin fragment of a sialic block is present in its foundation, and if so, whether it ought to be regarded as some remnant of a foundered continent or a sialic remnant of different origin [2]).

In spite of the above, there would seem to be no doubt that sialic blocks have foundered somewhere in the Pacific. A few of these masses were discussed in Chapter II. If we exclude such areas as Choco and the much-debated region of Burckhardtland, there would appear to be ample justification for the assumption that Cascadia represents a "border-land" of unknown extent, which probably acted as an area of denudation to the geosyncline of the Sierra Nevada up to Jurassic time, supplying it the while with detritus. It has probably sunk to a depth of 3,000 to 4,000 m below sea-level [3]).

In addition it should be noted that a land-mass of unknown dimensions has undoubtedly sunk into the Pacific along the coast of South-America [4]).

The assumption that Melanesia's eastern margin is a fault zone, on top of which formed the series of islands extending from New-Zealand to Samoa, would logically imply that a block must have foundered east of this area.

In fine, blocks of unknown extent may undoubtedly be said to have sunk in the marginal zones of the Pacific, but there can be no question that these constituted large sialic blocks, as enormous stretches of the floor of the Pacific appear to be built up of basic basaltic rocks.

The distribution of isobathic areas

The distribution of isohypses on the continents and isobathic contours in the oceans has led to interesting speculations. A graphic representation for the whole world, i.e. a so-called hypsographic curve of the earth's surface shows two frequency maxima, corresponding with two favored levels, one

---

1) A. Born, 1932, p. 767.

2) We refer the reader in this connection to the next paragraphs on submerged continents and to the final topic of this Chapter.

3) A summary of various data on Burckhardtland and a number of other regions appeared in my publication of the year 1937, see also he Appendix.

4) Gerth reported that a large stretch of mountain chains foundered near Payta, along the west coast of South-America (Reg. Geol. d. Erde, 1939, p. 2), and also writes that "ein Zweig des paläozoischen Gebirges sich nach N.W. in das Gebiet des heutigen pazifischen Ozeans fortsetzte" (Ibid. p. 6).

occurring at 100 m above sea-level, a second at a depth of 4,700 m below it. In other words the continental surfaces are situated at an average altitude of 5,000 meters above the ocean-floors. Wegener argued that the lower level should be regarded as the surface of the sima, whereas the upper level was that of the sial. These two levels of equilibrium would give rise to two frequency maxima, which would simply be controlled by Gauss' law of errors [1]). I wish to draw attention, however, to a few salient features of the submarine relief as known to us at present. The following table shows the distribution of isobathic areas in the Atlantic and the North Pacific Basin, according to Kossinna [2]). The areas are given in millions of square kilometers, the depth in km.

Depth	0,0–0,2	0,2–1,0	1–2	2–3	3–4	4–5	5–6	>6	Total area	Average depth in m
N. Atlantic Ocean	2,6	1,8	1,8	3,0	6,7	*11,5*	8,8	0,6	36,8	3,788
S. Atlantic Ocean	2,0	1,5	1,2	3,2	9,2	*15,2*	13,1	0,2	45,6	4,036
N. Pacific Ocean	1,1	0,8	0,8	1,9	8,8	21,2	*33,2*	3,0	70,8	4,753

The most frequent depth of the Atlantic appears to occur at the 4–5 km level, whereas that of the North-Pacific is generally one km deeper! [3]).

It seems to me that the explanation is obvious, viz. the deep level of the Pacific is controlled by the simatic nature of its bottom, the 1,000 m higher level of the Atlantic floor is due to the presence of a sial-sheet and the still higher level of the continents is the result of the fact that the latter consist of sial-flakes of a considerably greater thickness than the sial-sheet under the Atlantic.

We pointed out that the morphological features of the Atlantic and the Pacific are fundamentally different. If we glance at the new bathymetric chart of the Atlantic Ocean by Stocks and Wüst, it appears at once that the system of the Mid-Atlantic Rise and its adjoining ridges is situated at an average depth of 4 km or less (from 2–4,5 km), but the average level of the basins is 5 km (from 4,5–6 km)! Which one of these two is the original level of the Atlantic floor, the higher or the deeper one? If the basins are to be considered as secondary subsided formations, the primary surface of the Atlantic floor would be the higher level! Moreover, it seems very probable

[1]) It has to be admitted that a hypsographic curve for e.g. a jurassic or upper-paleozoic earth-surface would probably deviate in many, though not in fundamental respects from the present one. But I cannot endorse Bucher's criticism (1933, p. 52–59) that Wegener's interpretation of the two frequency maxima would be valueless.

[2]) Kossinna, 1933, p. 884.

[3]) In the words of Kossinna (1933, p. 885): "Nur ein grosses Meeresgebiet, der allerdings noch wenig erkundete Nordpazifische Ozean, weicht erheblich von den übrigen ab. Die 5000–6000 m Stufe nimmt hier allein 46,8 v. H. des Areals ein. Auf mehr als der Hälfte seiner Fläche (51 v. H.) ist der Nordpazifische Ozean über 5000 m tief. Die Flachsee und die Stufen von 200 m bis 3000 m haben dagegen eine geringere Ausdehnung als in allen anderen Ozeanen. Dieser Verteilung der Tiefenstufen entspricht die enorme Tiefe des Nordpazifischen Ozeans von 4753 m, welche die des Nordatlantischen um rund 1000 m übertrifft."

that the major positive relief-features of the Pacific basin proper, as compared to those of the Atlantic, are largely due to later accumulations of lava on the primary surface, which seems to lie at an average depth of 5,5 km below sea-level. These considerations stress the fundamental differences between the average depths of the Atlantic and Pacific ocean-floors. On the other hand, the rejuvenation of the relief is a factor which should be taken into account. Thus, the result which we arrive at here is that at least three distinct levels of primary importance may be observed. These are represented by the surface of the continents, the floor of the Atlantic and that of the Pacific. We will return to these features at the end of this chapter.

The problem

Now that the major characteristics of the suboceanic relief have been considered and before entering into a discussion of the various hypotheses which try to explain these features, the main points of the problem can be summarized as follows. (1) The "tetrahedral arrangement" of the continents, their antipodal relations to the oceanic receptacles and their peculiar shapes. (2) The occurrence of at least three favored levels on the earth's surface and their respective vertical distances. (3) The occurrence of a thin sial-sheet in the Atlantic sector and part of the Indian ocean. (4) The presence within these areas of a remarkable structural pattern dating from the early-Precambrian and the occurrence of certain tectonic features which have their counterpart on the adjacent continents. (5) The congruence of orogenic belts on either side of the Atlantic ocean, terminating abruptly on the coast. (6) The absence of a sial-sheet over the floor of the Pacific basin proper as well, probably, as in a few other oceanic areas mentioned above under groups 3 and 4. (7) The foundering of several sialic blocks within the area of the present oceans (Appalachia, Cascadia, etc.).

Continental drift

Alfred Wegener's much-debated hypothesis of continental drift seemed to give an elegant and simple solution of many problems of the earth's crust, but is it clear that his original conception (fig. 70 A) cannot be made to agree with our present knowledge of the floors of the Atlantic and Indian Oceans, for these are now generally believed to be sialic layers. Moreover, the structural pattern of these ocean-floors and their striking similarity to that of the surrounding continents cannot be explained by the drift-hypothesis. However, it might on the other hand be assumed that our continents occupied their present site as a result of sliding, which occurred in such a way that a sialic block of continental thickness was stretched. The stretched portion would then simultaneously have sunk and thus formed the bottom of the Atlantic and Indian Oceans (fig. 70 B). This deduction characterizes Du Toit's hypothesis of drift. In 1937 this same author wrote the following in a publication on the sub-marine relief of the Atlantic: "A pattern like this over so gigantic a region argues for a single controlling cause, and finds its proper answer only in continental sliding". To this he added a year later: "The relief of the Atlantic bottom shows all the characters of a stretch basin — particularly in the symmetrically-set, though crooked, Mid-Atlantic Rise with its lateral branches that reach out to either shore following north-easterly and north-westerly trends."

Yet I agree with Cloos that the striking similarity of the structures of the basins and surrounding ridges as compared to those of the adjacent continents is incompatible with the idea that these originated as a result of drift. Nor do I find it possible to reconcile any of the above conclusions as regards the Atlantic and Indian Oceans with the hypothesis of sliding continents, since the features of the bottom relief make it at all events clear that if the floors of said oceans originated as a result of stretching *this process must have occurred during the Precambrian.*

The above would be quite sufficient to reject the hypothesis of Du Toit and others, but I would like to examine another argument, which has recently appeared in several publications. Biological and paleontological data may be able to furnish more or less convincing evidence in favour of transoceanic land-bridges, but not — generally speaking — of the nature of such land-connections. We must make an exception for those data which seem to indicate that the distance between the continents had originally been a shorter one. Gerth, for example, argues that the similarity of the marine devonian strata of South-America and South-Africa pleads for a shorter distance between Africa and America during the Devonian. 50% of the lower-devonian species of the Falkland-Islands and 30% of the lower-devonian species of Brasil and Bolivia correspond with those of the South-African Bokkeveld beds. Gerth claims that this can only be explained if we assume that South-Africa lay nearer South-America in former times [1]). Still, these facts cannot be said to contribute any arguments in favour of the idea of continental drift. The same applies to the assertions of Gerth, Du Toit and others as regards a stratigraphic and structural similarity of the Gondwanides and Cape Mountains. It should be noted that the existence of analogous structures and stratigraphical sequences were reported by Holtedahl in such widely-separated regions as Spitsbergen and Scotland! (fig. 62). Yet no one would think of continental drift in this last case! Moreover, I wonder whether this hypothesis would be used to explain the following: the Maastrichtian fauna of Western Australia is closely allied to that of the Arigalus beds of the Trichinopoli district [2]) of India; and Spath [3]) recently described the Ammonites of Geraldton, Western Australia, and revealed that they closely resembled those of the Bajocian in England!

No other examples will be mentioned, for it would take us too far to discuss all the difficulties inherent to the hypothesis of drifting continents [4]) (we will deal more fully with some points of this hypothesis in Chapter VII). Besides, the conclusions arrived at in the preceding and following chapters

---

1) Gerth 1934, p. 66.

2) Teichert, Australian Journ. Sci. 2, 1939, p. 86. See also below (New Zealand), p. 165.

3) Journ. R. Soc. W. Australia, 25, 1939, p. 123.

4) For a more detailed examination of these difficulties see a publication of mine of 1937.

cannot be said to agree with the idea of an enormously extensive drift of continents *since the Paleozoic*. The rejection of the theory that the bottom of the Atlantic and Indian Ocean originated owing to drift during the

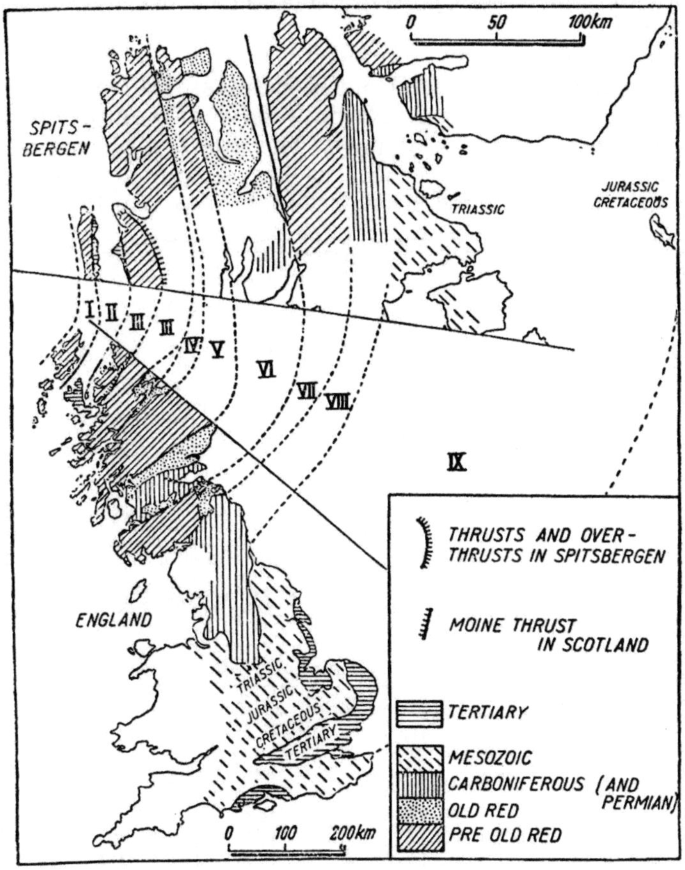

Fig. 62. A comparison between the geological structures of Scotland and those of northern Spitsbergen (After Holtedahl).

Paleozoic, or in even more recent times leaves us two rival hypotheses. The first is the idea of submerged continents, the second the hypothesis of permanence.

Submerged continents

Many geologists, following the example of Haug and Suess, have attached great importance to the reconstruction of large continents (fig. 63) which presumably sank to the bottom of the oceans.

Termier [1]), for instance, wrote: "....beaucoup de nos abimes océaniques

1) Termier, 1926, p. 359.

sont des gouffres relativement récents où se cachent des portions de l'ancien domaine continental; et si l'on pouvait vider ces gouffres de l'eau qu'ils contiennent, on verrait, au fond, des fragments de vieilles montagnes ou de vieux plateaux, qui se sont jadis étendus à la surface, sous la bienfaisante caresse du soleil".

Yet many geologists have refused to accept the idea of enormous submerged continents, as, apart from objections of a geophysical nature which will be discussed later, such a hypothesis would complicate the ocean's water-economy in a hopeless manner. The continents would (assuming that the volume of water of the oceans has remained constant) have to have been permanently flooded prior to the submersion of these huge blocks,

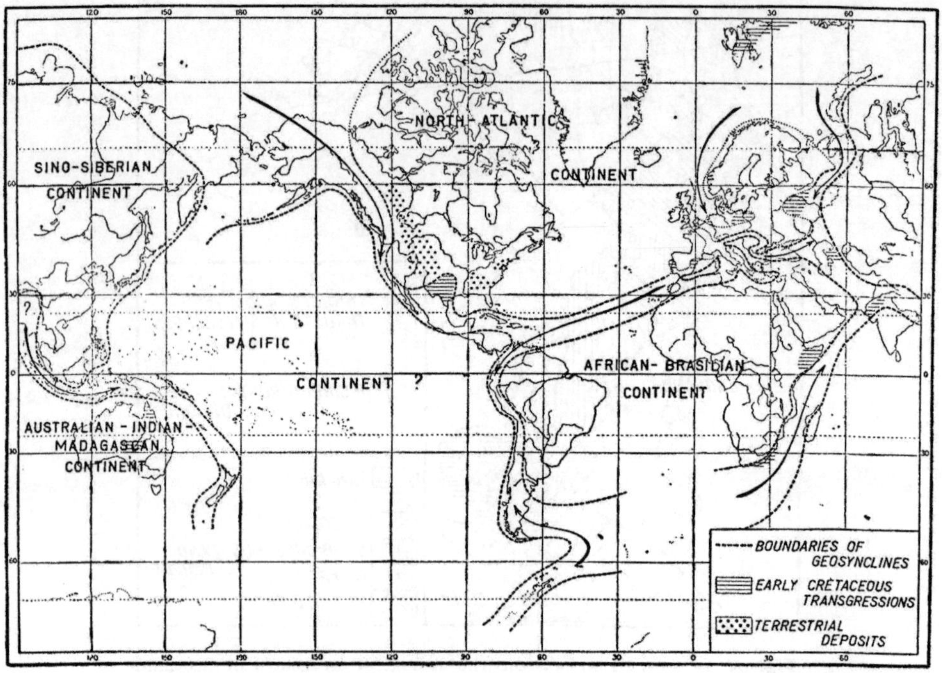

Fig. 63. Hypothetical reconstruction of transoceanic continents in the Lower-Cretaceous. (After E. Haug).

but this conflicts with all available data. The alternative would be that the volume of water would have increased steadily and to an enormous extent since the foundering of continental blocks. Yet this seems highly improbable. This point was clearly illustrated by Kuenen in the accompanying graph (fig. 64), which includes Walther's and Twenhofel's opinion, as well as his own. Kuenen writes in this connection [1]: "Walther held that there was probably no deep-sea until the end of the Paleozoic because no Paleozoic faunal elements have been found in the present deep-sea

[1] Kuenen, 1937, p. 460–462.

faunas. But this assumed absence of old deep-seas is only one of several possible causes that might explain what is after all a rather vague conception. It necessarily implies that the earth had an unimportant hydrosphere for some 1500 million years and then acquired its present volume of sea water at the rate of a few km³ per year or a few times the speed at which volcanic rocks are produced. Add to this that salts had to be furnished on the same gigantic scale and the impossibility of Walther's conclusion becomes manifest.

"Twenhofel's assumption, though less extreme, is still unacceptable. He says:

> 'In this inquiry it is assumed that pre-Cambrian and Paleozoic deep-seas existed, and that if the deep-seas of all time are considered they would be equivalent to deep-sea, with area like that of the present, extending as far back as the beginning of the Devonian'.

"In our graph, the volume of the oceans is plotted against geologic time, showing the views of Walther, Twenhofel and the writer. In Twenhofel's view the area abc must be equal to cde.

"There are many arguments for and against permanency of the oceans. In the opinion of the writer the most forcible are: the absence of fossil deep-sea sediments from the continents and the impossibility of sudden

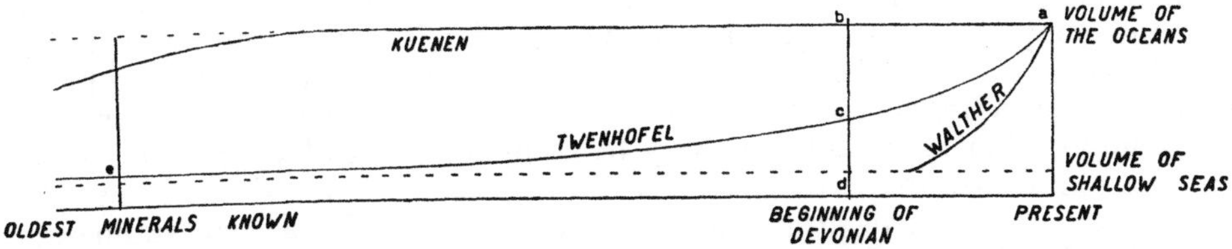

Fig. 64. Total amount of oceanic waters in geological time according to the views of Walther, Twenhofel and Kuenen (After Ph. H. Kuenen).

production of water and salts since the juvenile stage of the earth. For post-Algonkian times doubt as to the existence of permanent ocean basins is quite unreasonable; although their position and shapes may have varied through continental drift and although isthmian links may have arisen and disappeared".

This question is closely related to that of the total amount of sedimentation in the deep-sea and to the whole complex of geochemical processes of the continents and oceans. So far, however, there seems to be no evidence of an annual increase of a few km³ in the amount of the oceanic waters, and it is quite as improbable that the salinity increased at the same rapid rate. Walther's curve would therefore appear to be too extreme, and — unlike Du Toit, whose point of view is expressed in a publication of 1940 on the subject of the Pacific Ocean — I am of the opinion that Kuenen's graphic representation is probably more in accordance with the facts than Twenhofel's curve.

Two conflicting deductions are found side by side if we adhere to the hypothesis of submersion. The first asserts that vast continental blocks foundered in the course of time. The second claims not only that the total amount of the oceanic waters has not increased proportionately since the Cambrian, but that it has even remained almost constant. Should one nevertheless cling to the theory of submerged continents, the only alternative would be to assume that while vast blocks were being submerged in one area, parts of the ocean-floor of an almost identical size were being elevated in others, i.e. — to cite one example — that the bottom of the Pacific, which was originally situated at a far greater depth, rose while blocks sank into the Atlantic and Indian Oceans, and that the total amount of water and its surface level consequently remained almost wholly unaltered. These considerations are indicated by the line B-C in the diagram in fig. 65. The line A-C, on the other hand, illustrates the idea of a relatively stable floor of the Pacific. In addition to this, both graphs show the periodic cadence of the floor of the Pacific, as deduced from the rhythm of transgressions and regressions.

It is not quite clear, however, why such opposed movements should have occurred in areas of almost equal extent. Nor is it clear why these movements should have occured in such a way that the sea-level remained comparatively stable. Besides, the above assumption gives rise to further complications, for if the phenomena can be attributed to this process, it might be asked what happened in the substratum. To begin with, it remains entirely problematical (in spite of Barrell's speculative though interesting reflections on this subject [1]) why a given area such as, for instance, the present floor of the Atlantic should have been submerged, while the neighbouring blocks remained standing as continents. Moreover, this process involves more than just submersion, and cannot merely be compared to a block of wood which is pressed down more deeply into the water. Seismic data show that the blocks must have grown much thinner during the act of submersion (fig. 70 C). This might be attributed either to stretching, or else to the melting of sialic material from the lower part of the crust, which subsequently vanished. The first alternative was discussed in the preceding paragraph. This stretching-process might possibly have occurred during an early stage of the earth's history, but contributes nothing towards the solution of the problems which have arisen as regards the Cambrian and later times, as the features of the sea-bottom (which were dealt with extensively above) appear to contradict the idea of a more recent process of stretching. The second contingency would only force us to accept some additional *hypothesis ad hoc*, with no sound basis whatever.

The above clearly shows that the hypothesis of submerged continents

---

[1]) Barrell made some interesting reflections on the possible cause of the subsiding movement, suggesting that the foundering might have been the result of "the weight of magmas of high specific gravity rising widely and in enormous volume from a deep core of greater density into these portions of an originally lighter crust". Barrell claims that this regional foundering occurred especially in primordial times (though it did not cease altogether then, owing to the periodic accumulation of heat from radioactive processes deep beneath the crust), and draws a comparison with the lava plains of the moon. Yet this hypothesis, too, fails to explain the presence of an upper sialic layer of the Atlantic type, or the absence of such a layer in the Pacific basin proper, and the same applies to the comparatively important thickness of the continental bucklers.

leaves many points unsolved, and we consequently proceed to the third hypothesis with the unpleasant sensation of having been twice disappointed.

The hypothesis of permanence

The aspect of the continents and oceans has not escaped frequent alterations. We need only think in this connection of the East-Indian basins, Melanesia and Appalachia. Some parts have been submerged, even though such regions as Cascadia, Appalachia and Scandia were no larger than "border-lands", as Schuchert called them. The idea of permanence is consequently untenable in its extreme sense. But just as Schuchert and Willis proclaimed themselves advocates of the theory of permanence, in spite of the border-lands and isthmian links which they themselves reconstructed, so, too, we can speak *cum grano salis* of permanence as regards the oceans as long as we refrain from reconstructing huge submerged continents (fig. 66).

This hypothesis will have to be based on three premisses: (1) that comparatively thin sialic layers existed in the floors of the Atlantic and the Indian Ocean since at least Cambrian time; (2) that in spite of this, relief-features such as basins and ridges resembling those of the intervening African continent originated within these blocks, and that these were brought about by the same subcrustal processes; and (3) that the amount of oceanic waters has not changed to any important degree since the Cambrian.

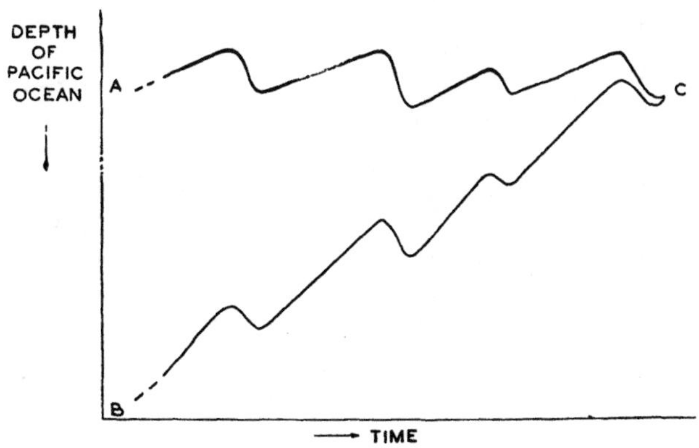

Fig. 65. Schematic diagram illustrating the ideas of a relatively stable (A—C) and a rising floor (B—C) of the Pacific Ocean.

Among the features which are liable to change in one or perhaps several respects are the details of the submarine relief. For example, here too, we need only think of such foundered blocks as Appalachia, the possibility of widely diverging periods of origin of the basins of the Atlantic and Indian Oceans with their possibly later movements, and the "rejuvenation" of the relief, which is so strikingly illustrated by the Mid-Atlantic Rise and the Carlsberg System. These movements remind one at once of the "isthmian links" of Schuchert and Willis (fig. 67). Reflections of a highly interesting though speculative nature were made by Willis and Nölke on the emersion and submersion of narrow transoceanic isthmi. Their origin and submersion will probably remain a mystery for some time to come, but the inference that such features had in fact been created and then vanished would not seem to be an improbable one. Cloos [1]), too, when dealing with the analogy of the characteristics of the continental and oceanic sectors (Atlantic and Indian) drew attention to the fact that the narrow zones surrounding the continental basins can look back upon a particularly agitated history, with important vertical upward and downward oscillations. I consequently incline towards the view that no essential objection can prevent us from accepting the idea of isthmian links as postulated by Schuchert and Willis.

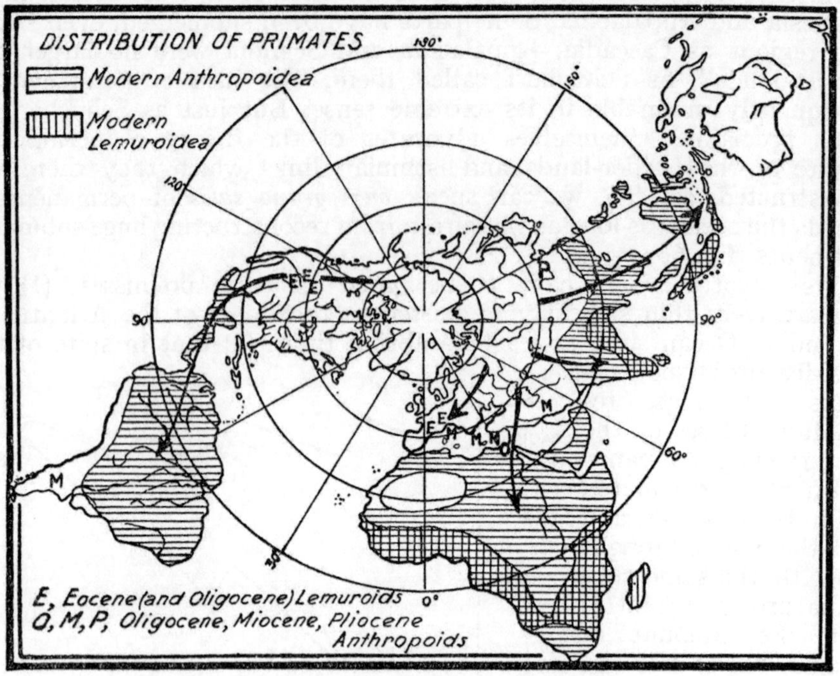

Fig. 66. The distribution of Primates according to the hypothesis of permanence. (After W. D. Matthew).

Of course, it cannot be denied that the existence of land-connections during certain given periods cannot be proven geologically. Each so-called reconstruction of a transoceanic land-bridge must necessarily retain a hypothetical character. Yet such land-bridges need not be discarded *a priori* as mere products of the imagination.

Another changeable factor is the periodical variation of the distance between the floor of the sea and the surface of the continents. This question was discussed in Chapter V.

A few fundamental questions still have to be solved, however, viz. the way in which the relatively thin and deeply situated sialic blocks existing since the Cambrian came to be formed, what caused the Pacific floor to be sial-free, and how the antipodal arrangement of continents and ocean-floors came into existence.

---

1) Cloos, 1937, p. 344.

In attacking these fundamental problems the birth of the moon will once more have to be considered. If we accept the resonance theory of the moon's origin it seems highly probable that a great part of the earth's outer shells entered the moon's formation. This process may account for the remarkable differences between lunar and terrestrial volcanism, as suggested by Escher (see p. 8). But what happened to the remaining sialic matter of the earth? There are two possibilities, viz. either the terrestrial remnants of sial were able to flow out and unite again as one continuous shell, enveloping the entire world, or they were unable to do so, being too rigid. The last possibility is basic to the theories of many authors. In their opinion the earth was originally enveloped by a sialic shell of continental thickness. After the moon's disruption the remaining sial-fragments of the earth would have formed our continents, floating on the heavier simatic substratum that forms the bottom of the oceanic receptacles (fig. 70 A). More than once the question was raised whether the Pacific might not be considered as a scar brought about by the separation of the moon. Our continents should be compared to the acid slags floating on the surface of the heavier melt in a blast furnace.

Practically the same idea can be found in the theories of Osmond Fisher, Pickering, Taylor, Wegener, Schwinner and Escher, and the last author even tried to calculate the thickness of the lunar sial by computing the mass of terrestrial sial that originally filled the gaps of the oceanic sectors.

It is at all events clear, however, that neither the continents nor even their so-called nuclei (representing the continents' innermost structures) should be regarded as undisturbed remnants of that distant and turbulent period in our planet's infancy. The preceding chapters showed that these "acid slags" had from the early beginning a very complicated history and that the continents (parts of which have been breaking down until quite recently) formerly occupied a much more extensive area than they do at present. We know, moreover, that the history of the oceans' submarine relief is a far from simple one. Another point is that the bottom of the Atlantic and Indian Oceans differ considerably from the floor of the Pacific.

One question of primary importance, however, is: *why should the sialic material have flowed out to form a continuous shell on the moon and not on earth?* The following points should not be forgotten in this connection. It is questionable, but at any rate quite possible that prior to the disruption the sialic material had — it is true — differentiated into the outer parts of our globe, but that it was not yet in a solid state. Every one will admit (see fig. 4) that the outer shells of the earth were of such a fluid composition that they readily responded to the tidal forces which deformed the earth's geoid before as well as after the moon's separation. It will also be admitted that part of these outer shells shifted towards the moon and that the scar of the separation was closed smoothely. Moreover, there does not seem to be the slightest doubt that when the earth, recovering from its amputation, was moulded into its new shape, the simatic layers — even the uppermost — formed a fluid mass uniting into a continuous shell. It seems, therefore, quite unreasonable to suppose that the thin upper layer of sial (see fig. 5) would have been the only one incapable of reacting in the same way [1]). Furthermore, though the hypothesis may be assumed to account for a few "acid slags" floating in the basic substratum that constituted the floor of the oceans, it cannot explain the origin of a thin sial layer such as that of the Atlantic. It might eventually explain *two* favored levels of the earth's surface, but no more than that. The following pages will also draw attention to some of the reasons for the assumption that a rigid crust, incorporating the sial flakes, only solidified during a later part of the earth's history.

The above points may be said to constitute a few objections as regards the hypothesis of the origin of continental bucklers and oceanic receptacles and would seem to show that this view is open to serious doubt.

The second contingency to which reference was made above will now be examined. This possibility, remarkably enough, has never been followed up to the end by anyone, so far as I know. A new hypothesis will consequently have to be formulated.

A new hypothesis

The preceding considerations are based on the resonance theory of the moon's origin and are valueless if we accept the opposite view, viz. that the moon and the earth formed simultaneously as two separate bodies (see Chapter I).

The new conception, however, has the advantage that it is quite independent of any other hypothesis with regard to the origin of the moon. Our considerations will in this case have to be based on the supposition that the earth was initially enveloped by a continuous sialic layer, gradually solidifying and floating on a denser basic substratum [2]). It may be that this layer formed after the moon's disruption, but it might similarly be assumed that the moon had not in fact been severed from the earth.

The only possible way in which continental blocks might have originated from a world-encircling sialic layer, growing gradually cooler and solidifying, is that a process of thickening occurred, resulting from folding and drifting [3]). This process shortens the sialic layer, and sial-free parts (such as large portions of the present Pacific area, with the possible inclusion of the oceanic bottoms mentioned in groups 3 and 4, or parts thereof, see page 82) were therefore formed.

The diagram in fig. 68 shows how a continent may possibly have originated through periodical buckling and drifting of an originally thin sialic layer

---

1) Even the lighter simatic material would have been lacking in the Pacific sector according to Schwinner (1935, p. 316) and this would have prevented the sialic layer from covering this area. The fact, however, that the Pacific's volcanoes are built up of basaltic material clearly shows that this opinion can not be upheld.

2) cf. Daly 1938, p. 176.

3) This is the same idea, after all, as that met with previously when retracing the earth's history from its more recent stages to the remote ages of its very beginnings (see Chapter IV). Since the oldest continental shields are known to consist of intensely folded belts, the question naturally arises whether the continents were not the result of a process of thickening, i.e. of a periodical process of buckling of an originally thinner and world-encompassing sialic layer. A tentative and questioning reference was made to said considerations in Chapter IV.

enveloping the whole earth. It is obvious that this process must have attained its most important effect in the early Precambrian, as the whole later process of buckling and folding of the continents is known to have only evolved in these existing precambrian bucklers [1]).

This is indeed remarkable and will have to be examined before any other features are discussed.

In the beginning the continents grew steadily larger, but this has not been

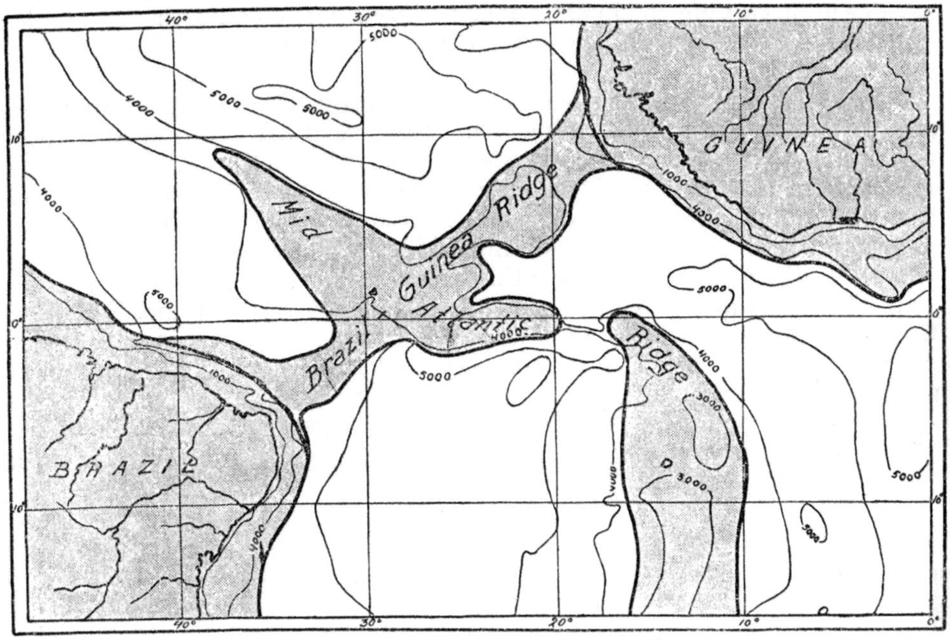

Fig. 67. A transatlantic isthmian link (After B. Willis).

repeated since at least the Cambrian or probably even earlier times, as the geosynclines and folded chains that are known to have formed since then originated in a basement which had already been folded previously [2]). Hence we cannot escape the conclusion that some event of fundamental importance must have occurred during the early part of the earth's history, causing the surface history of primeval times to differ from that of subsequent periods. By this we mean that from a certain critical moment onward continental growth ceased.

The most plausible explanation seems to be that this event of fundamental importance was the formation, or better still the completion of a world-encircling solid crust, which had become so thick that subsequent growth

1) Even on those rare occasions in which abyssal sediments have been observed, it seems unlikely that these areas should be regarded as exceptions. This is at the most doubtful.

2) A fuller illustration of the preceding consideration will be found in Chapter II and the notes on Plates 1 to 5 in the Appendix.

was no longer possible. Such youthful caprices as are illustrated in fig. 68 may be expected to have ended as soon as the earth was enveloped by a solid crust and attained a physical state more or less resembling that which it has at present. From that time onward the terrestrial forces were imprisoned and became *subcrustal* processes.

At that time the crust consisted of basic rocks in the sial-free parts of the surface, just as it does now, and it may have contained both sialic and simatic rocks in those regions of the world where sial-flakes were present. It must have been of varying thickness, though it need not of course have equalled its present-day thickness. We will call this state the "consolidated surface of the earth".

Prior to the formation of a world-encircling crust, the terrestrial dynamics probably evolved with the same periodicity as later on. Yet the effect produced thereby on a non-consolidated surface was apparently a very different one than that produced on a solid crust by the "deeply imprisoned titans". Before the formation of a consolidated surface, the sialic upper layer was folded into continental thickenings. This was the period of continental growth. With the solidification of the earth's surface this process came to an end. The question as regards the possible cause of the periodicity [1]) of terrestrial dynamics will not be discussed for the moment. It forms a separate problem, to which attention will be paid towards the end of Chapters VII and IX (p. 127, 141 and 155).

If the above hypothesis may be said to contain a germ of truth, the problem of the continents and ocean-floors would consequently be associated with

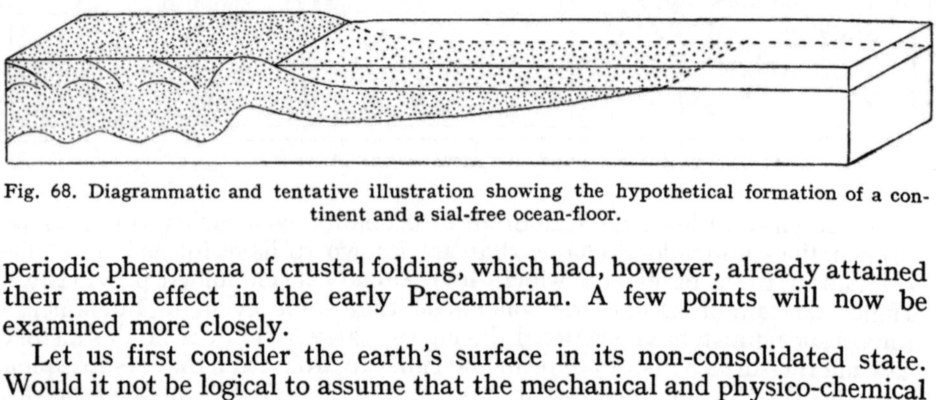

Fig. 68. Diagrammatic and tentative illustration showing the hypothetical formation of a continent and a sial-free ocean-floor.

periodic phenomena of crustal folding, which had, however, already attained their main effect in the early Precambrian. A few points will now be examined more closely.

Let us first consider the earth's surface in its non-consolidated state. Would it not be logical to assume that the mechanical and physico-chemical processes of the earth elapsed in such a way that a certain equilibrium as regards the distribution of lighter and heavier masses in its interior and on its

[1]) The periodical processes that act on the earth's surface must at one time have found the sial-sheet cooled and solidified to such a degree that the dynamics referred to above were able to begin their activities. It may readily be assumed that the primordial sialic surface layer was not so purely homogeneous that it did not yield more easily in some places than in others, thus producing the very first crustal thickenings. I do not believe that there are any data at present that would enable us to ascertain the exact time of this event, and the same applies to the formation of the consolidated surface.

surface resulted there-from? Might we not therefore suppose that the specific light crustal bucklers (we will refer to them as ,,primordial continents") went adrift, assuming a remarkable pattern marked by an antipodal relation between continental and non-continental i.e. the later oceanic sectors. This suggestion gives no adequate explanation of the actual shapes and location of our continents, but I believe that we may assume some primeval process of drifting to be responsible for the remarkable tetrahedroid pattern of our globe.

In several of the preceding paragraphs (p. 96, p. 100) it was pointed out that the Atlantic floor might have formed as a result of stretching in the sense Bucher and Du Toit attached to this process, but special emphasis was laid on the fact that in that case it would have occured at any rate in the Precambrian, as morphological and geological arguments showed above. It may be that during the process of drifting of these primordial continents a sial-sheet was torn into two or more parts, each moving in a different direction and producing a thinly stretched sial-flake between them. In this way (and only during these primordial times) a thin sial-sheet like that which probably exists under the Atlantic may possibly have originated as a result of stretching of either a primordial continent or the intervening primary sial-layer. There is reason to believe that this process actually happened. For if the sial of our present continents were imagined to spread out as thickly as the sial-layer of the Atlantic, it would probably cover a larger area than that of the present sial-free ocean bottoms. A floor such as that of the Atlantic cannot therefore be regarded as a remnant of the primary world-encircling sialic shell. On the contrary. It would seem to have grown thinner as a result of stretching during the early Precambrian. It would be impossible to say just now whether a fragment of the primary sialic layer of original thickness exists in some part of the world to-day.

Let us now consider the moment that these primeval surface-features became petrified, as it were, by the formation of a world-encircling solid crust. Among the first topics to be mentioned in this chapter was the noteworthy structural pattern of the floor of the Atlantic and the surrounding continents. It was shown that this pattern had probably originated in the early Precambrian and attention was drawn to the part which these old structural lines repeatedly played in later times (p. 84–86). We arrived at the same conclusion when dealing with the continental basins (p. 47). It might therefore be asked whether it is not obvious that this ancient pattern dates from the period of formation (i.e. from the consolidation) of a solid crust.

In the preceding paragraphs we came to the conclusion that three favored levels occur on earth, represented by (1) the continental surface, (2) the floor of the Atlantic, and (3) the bottom of the Pacific Ocean. It need hardly be said that these results fit in well with our working-hypothesis. The difference in the level of the primeval floors of the Pacific and Atlantic is seemingly due to the difference of the materials composing either floor. On the other hand, buckling of the primordial sialic crust was responsible for the continental thickenings. Their remarkable thickness and high level, as compared to that of the bottom of the Atlantic, is controlled by five factors, viz. (1) the thickness of the original world-wide sialic layer, (2) the compressive forces during the periodic phases of early precambrian buckling, (3) the leveling effect of subaeral erosion, (4) the precambrian stretching of sial-sheets of the Atlantic type, and (5) the local thickening of the continental basement-complex as a result of the formation of later geosynclines and mountain-belts.

The above considerations indicate the main outlines of the new working-hypothesis. I do not wish to suggest that this hypothesis solves the intricate problem of the ocean-floors. I merely followed a line of thought which had hitherto been neglected, and as all the previous hypotheses are deficient in many respects, the new "working hypothesis" may stimulate others in their efforts to unravel the mystery of the ocean-floors. I fully realize how many difficulties are inherent to this new aspect of the old problem, and will immediately discuss some of its more complicated elements.

I would first like to draw attention to a difficulty, however, which is not only attached to the working-hypothesis, but which may be said to characterize every other kind of hypothesis in which a certain degree of permanence is assumed in oceanic floors of the Atlantic type. For how are we to explain the abruptly ending Caledonian and younger folded mountains on either side of the ocean? Stratigraphy and tectonic data show that a strip of unknown extent is lacking. Yet the morphology of the ocean's floor contains no indication of submerged Caledonian or other folded chains. Moreover, Haug remarked quite rightly that the theory of transoceanic mountain-chains requires, amongst other things, that a transoceanic geosyncline, including an area of erosion on at least one of its sides, should have existed in this area prior to said formations. It was this that induced Haug to reconstruct those continents, the existence of which we rejected on the grounds mentioned above. We can therefore easily understand that Stille doubted the former existence of transatlantic mountains [1]), and fig. 69 shows how Stille imagined the Caledonian and Variscian Mountains to have bent back originally towards the continents (see also Plates 1 and 2). Yet even so, another fact remains, viz. that the abruptly ending belts are situated approximately opposite one another on either side of the ocean. The following suggestions may help to overcome these difficulties. Let us, for example, suppose that, owing to the influence of

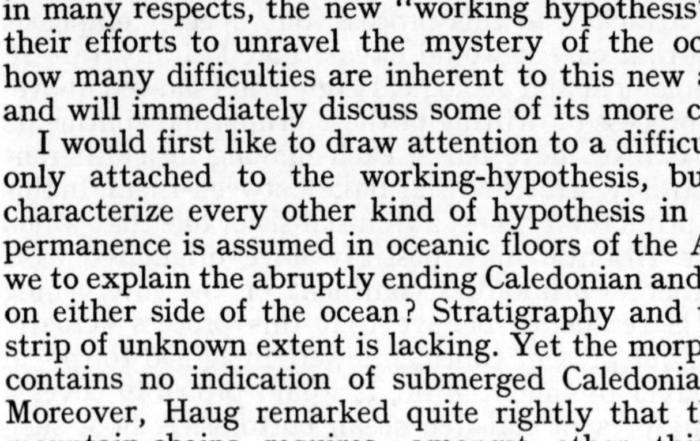

Fig. 69. A hypothetical reconstruction of non-transatlantic mountain-chains (After H. Stille).

[1]) Stille, 1939, p. 343, 345.

subcrustal processes, the earth's crust had reached a stage where it became possible for a geosyncline to form within a certain zone. Let us also assume that this zone continued under an area such as that of the Atlantic. In that case a geosyncline and folded chain would potentially be able to originate within a certain zone of the floor of the Atlantic. Yet, such a thing would never happen, for insufficient erosion products would be available for the purpose of filling the furrow. Should a subsiding trough originate within this zone and be buckled, the result would be a sialic root of considerably smaller dimensions, for the sialic layer of the Atlantic is a much thinner one than that of a continent. Should this nevertheless happen, the isostatic anomaly would be much smaller than that in the contemporaneous extension of this zone on the continents (it might besides be questioned whether anything like a geosyncline would ever originate in a suboceanic sialic layer). This would help to show that a transoceanic connection need never have existed above sea-level, and that the stratigraphy, tectonic structure and epochs of folding of mountain-belts are more or less analogous on either side of the ocean in spite of the absence of transoceanic mountain-belts. It would then no longer be necessary to look for the formation and subsequent disappearance of a missing link which had never been visible above sea-level — nay, which had never even existed.

Moreover, the above considerations show a way in which to test the hypothesis of permanence. For if transoceanic chains foundered in the ocean, gravimetrical observations might be expected to reveal them as a strip of distinct deviation of the isostatic equilibrium, occuring in the prolongation of the abruptly ending zones of folding on the continents and presenting anomalies of the same intensity as those in this particular continental belt. On the other hand, it would be more in accordance with the hypothesis if these anomalies were non-existent or much fainter than those in the adjoining continental strips. So far as I know, gravimetrical observations at sea are not numerous enough and far too wide-spread at present to enable us to test this surmise. The same applies to a possible testing of sunken border-lands and isthmian links.

To return to the working-hypothesis — we might add the following. It was shown above that terrestrial dynamics caused a similar pattern to form in both the continental and the oceanic sial-sheets during the consolidation of the surface. It should therefore not be wondered at that both areas responded in an identical fashion to the subcrustal forces of later times. Hence phenomena such as the formation of basins and furrows, graben and dome-shaped elevations and the rejuvenation and foundering of blocks obviously occurred in both these sectors. It should be noted, however, that there is one phenomenon which would never manifest itself in oceanic areas, viz. the formation of transoceanic mountain-chains. The only factor to prevent the formation of the latter is the absence of detritus resulting from atmospheric denudation.

It might be objected that no attempt has been made to explain the foundering of border-lands or the sinking of certain oceanic basins. These phenomena, however, occur in quite the same way on the continents, and the problems arising from their origin and history are not confined to oceanic sectors. The submergence of border-lands, for instance, cannot in itself be regarded as an unusual phenomenon. For it should be noted that blocks which were first submerged, then elevated, and then once more submerged and elevated, are also met with on the continents. The sub-oceanic features and the similar continental characteristics cannot be explained at present, for our knowledge of precambrian history and terrestrial dynamics is not yet extensive enough. Hence the fact that no adequate explanation can be given of these features does not imply that the above working-hypothesis is a deficient one.

I would like to elucidate a few remaining points. The hypothesis to which we referred above is based on the assumption that a continuous sialic layer, which was originally enveloped by a thin layer of water, extended over the entire world. Oceanic basins with a steadily increasing volume of water originated *ipso facto* as the continental blocks were created. It would consequently be wrong to presume that the continents emerged from the deep-sea, and that we might therefore expect to find the continents covered by a veneer of deep-sea sediments. On the contrary, deep-sea deposits are known to be almost non-existent in the present continents. Yet this cannot be said to conflict with our working-hypothesis, the more so as the volume of water has undoubtedly grown since primeval times (see fig. 64, where we find this opinion expressed even in the top curve!).

As pointed out in Chapter II, the whole history of geosynclines and folded chains since the Cambrian took place on a basement which had already been folded during the Precambrian. It is impossible to describe the subcrustal processes which were responsible for the distribution and sequence of folding on the continents (as shown on plates 1 to 5). It is evident, however, that a repeated buckling of continental areas — a process which may possibly also have taken place in the Precambrian — must have caused the lower side of the continents to become extremely irregular. This feature is in fact corroborated by both seismic data and the different altitudes of the mountain-belts, which are more or less in a state of isostatic equilibrium.

Summary

Four major morphological units can be observed in the relief of the ocean-floors. They are: (1) the Atlantic and the western part of the Indian Ocean; these areas are characterized by numerous basins with comparatively flat bottoms, separated by relatively narrow and steep ridges; (2) the large North Pacific basin. The most prominent characteristics of the latter are the diagonally-arranged ridges; (3) the eastern part of the Indian Ocean, the south-western part of the Pacific, and the three Antarctic basins, all of which are marked by but a slight subdivision of the relief; (4) the North Polar basin, which would appear to concur in many respects with the basins of the third group. We will not consider the areas mentioned sub (3) and (4), for data on these regions are still very scarce.

An examination of the Atlantic and Indian Oceans, and a comparison with the available geological and seismic data, shows that the bottom probably consists of sialic material. Internal forces caused similar events to occur in this sialic layer and the surrounding continents (e.g. basins and ridges were formed, and eventually rift-valleys). The basins and ridges have a structural pattern and symmetrical arrangement which seem to date from the early Precambrian, but the relief of the ocean-floor underwent the same rejuvenation as the continents in comparatively recent times (typical examples are the Mid-Atlantic Rise and the Carlsberg Ridge). There can hardly be any doubt that some blocks of as yet unknown extent (such as Scandia, Appalachia and the Arabian Sea) have become submerged.

The arrangement of the linear ridges in the Pacific, crowned with many volcanoes, is attributed to the magma's finding its way upward along faults and fissures in the ocean-floor. This floor of the Pacific appears to be built up of a rigid crystalline layer of basic rocks (crystalline sima). The presence of nepheline-free lavas in the area of the Albatross Plateau may perhaps be attributed to the existence of a thin sheet of sial in its foundation. Sialic blocks seem to have foundered in the marginal areas around the Pacific (e.g. Scandia), but the fact that the major part of the floor of the Pacific is composed of crystalline sima conflicts with the idea of a submerged Pacific continent.

The most frequent depth of the Atlantic appears to occur at a level of 4–5 km, whereas the North-Pacific is generally one km deeper. The explanation seems obvious, viz. the deep level of the Pacific is controlled by the simatic nature of its bottom, the 1,000 m higher level of the Atlantic floor is brought about by the presence of a sial-sheet and the still higher level of the continents is the result of the latter's composition of sial-sheets, which are considerably thicker than the sial-flake under the Atlantic.

The available morphological, geological and geophysical data cannot be stated to concur with either the hypothesis of a large continental drift in comparatively recent times (fig. 70 A), or with the hypothetical assumption that the floors of the Atlantic and the Indian Ocean originated in the Paleozoic or at an even more recent date as a result of stretching of continental blocks (fig. 70 B).

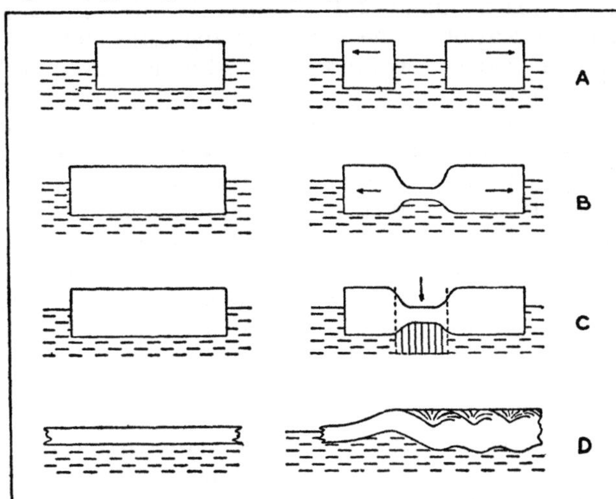

Fig. 70. Schematic representation of four different views on the origin of ocean floors.

The hypothesis of large submerged blocks does not agree with the water-economy of the oceans (fig. 70 C). For if the volume of water of the oceans is assumed to have remained relatively stable, the continents would have to been permanently flooded prior to the foundering of the continental blocks, and this conflicts with all that is known of its geological history.

The alternative, viz. that the volume of the oceanic waters has increased by a few km³ per year is equally improbable. The only way out of the impasse would be to suppose that, while large blocks were sinking in one area, other equally extensive blocks were being raised elsewhere to the same extent. Such a phenomenon would in itself be unacceptable, but apart from this objection it would be difficult to understand why certain continental blocks should have submerged to the exclusion of others. The last point to be noted is that submersion itself does not only give rise to insurmountable objections, but that the thinning of the gigantic blocks forms an additional difficulty. Many fundamental difficulties will therefore have to be solved if we are to explain the problem along the lines of the hypothesis of submerged continents.

The hypothesis of permanence should be taken *cum grano salis,* for sialic blocks of restricted size are known to have foundered as "border-lands". If the bottom of the Atlantic and Indian Oceans may be assumed to have existed as thin sialic layers since at least the Paleozoic, and the total amount of oceanic waters to have remained almost constant since then, two changeable factors remain. The first is represented by the details of the submarine relief, the second by the periodic changes in the distance between the bottom of the ocean and the surface of the continents. The first includes diverging periods of formation of (and later movements within) the basins of the Indian and Atlantic Oceans, as well as a rejuvenation of the relief (e.g. the Mid-Atlantic Rise and the Carlsberg Ridge), and the formation and submersion of isthmian links. The origin of these movements must necessarily remain problematical, but the similarity of the features of the oceanic and continental areas hardly leave any doubt that they might have occurred. This brings us to the fundamental potentiality of isthmus-shaped transoceanic land-communications. The second changeable factor is the pulse of the oceanic floor as derived from the world-wide trans- and regressions described in Chapter V.

However, the following major problems remain to be solved, viz. the comparatively thin sialic layer covering the floor of the Atlantic and Indian Oceans, the absence of such a layer in the Pacific, the antipodal arrangement of continents and oceanic depressions, and the abruptly terminating orogenic belts on the coasts of the Atlantic.

Many authors assert that the earth originally was enveloped by a sialic shell of continental thickness. If we accept the resonance theory of the moon's origin, it seems highly probable that a great deal of the outer shells of the earth entered the formation of the moon, and the remaining terrestrial sial-sheets would have formed our continents, floating on the heavier substratum that constituted the bottom of the oceanic depressions (fig. 70 A). But why would the sialic material have formed a continuous shell on the moon and not on earth? Besides several objections, showing that this point of view is open to serious doubt, were raised against this hypothesis in the preceding paragraphs.

The only alternative — leading to a new working-hypothesis — appears to be that the earth had initially been enveloped by a continuous sialic layer which was gradually solidified and floated on a denser basic substratum. It may be that this layer originated after the moon's disruption, but it might similarly be supposed that the moon was not severed from the earth.

The only possible way in which continental blocks might have originated from a world- encircling sialic layer is that the latter was thickened by folding and drifting. The sialic layer is shortened by this process, and sial-free parts, such as the present North Pacific basin, were thus formed (fig. 70 D). To restore the equilibrium of the distribution of lighter and heavier masses on its surface, the specific light crustal bucklers went adrift and their arrangement assumed the remarkable pattern known as the antipodal relation of continental and oceanic sectors, which simulates a tetrahedroidal pattern. During the process of drifting, a thin sial-sheet, — such as may probably be said to exist beneath the Atlantic —, may have originated as a result of stretching of either a primordial continent or the intervening primary sial-layer. The occurrence of three favored levels on earth agree with this surmise. An event of fundamental importance was the completion of a world-encircling solid crust, which had gradually become so thick that subsequent continental growth was no longer possible.

Moreover, it would seem plausible to assume that the curious structural pattern which characterizes both the floor of the Atlantic and the surrounding continents and which dates from early Precambrian times came into existence during the above process of crustal consolidation.

The trans-Atlantic similarities may possibly be explained by the suggestion that if a geosynclinal zone continued under an area like the Atlantic, no mountain-belt could have originated in the oceanic sector, since no area of denudation and therefore no detritus was available. Hence mountain-belts on either side of the Atlantic may concur as regards their respective stratigraphy, epochs of folding and tectonic structure in spite of there never having been a transoceanic connection above sea-level. We will be able to test this idea as soon as gravimetric observations at sea have become more numerous.

References

BANDY, M. C. *Geology and petrology of Easter Island* (Bull. Geol. Soc. America 48, 1937).

BARRELL, J. *The origin of ocean basins* (pp. 39–43 of Chapter I, in "The Evolution of the earth" 1919).

BARRELL, J. *On continental fragmentation, and the geologic bearing of the moon's surficial features* (Americ. Journal of Sci. 13, 1927).

BUCHER, W. H. *The deformation of the earth's crust* (1933).

BUCHER, W. H. *Submarine valleys and related geologic problems of the North Atlantic* (Bull. Geol. Soc. America 51, 1940).

BULLARD, E. C. and GASKELL, T. F. *Seismic methods in submarine geology* (Nature 142, 1938)

CHUBB, L. J. *The structure of the Pacific basin* (Geolog. Magaz. 71, 1934).

CLOOS, H. *Zur Gross-tektonik Hochafrikas und seiner Umgebung* (Geol. Rundschau, 28, 1937).

DALY, R. A. *Architecture of the Earth* (1938).

DU TOIT, A. L. *Our wandering continents* (1937).

DU TOIT, A. L. *The origin of the Atlantic-Arctic Ocean* (Geol. Rundschau 30, 1938).

DU TOIT, A. L. *Observations on the evolution of the Pacific Ocean* (Proceed. Sixth Pacific Sci. Congr. California 1939, vol. I, 1940).

EWING, M. ,MILLER, B. L. a.o. *Geophysical investigations in the emerged and submerged Atlantic coastal plain* (Bull. Geol. Soc. America 48, 1937).

FARQUHARSON, W. I. *Topography* (John Murray Exped. Rep. I, Nr. 2, 1936).

FIELD, R. M. *Structure of continents and ocean basins* (Journ. Washington Acad. Sci. 27, 1937).

FIELD, R. M. and JONES, D. M. *Geophysical explorations of ocean-basins* (Quart. Journ. Geol. Soc. London 1938).

FURON, R. *La Paléogéographie* (1941).
GERMAIN, L. *Le problème de l'Atlantide et la zoologie* (Ann. Geogr. 22, 1913).
GERMAIN, L., JOUBIN, L. et LE DANOIS, E. *Une esquisse du passé de l'Atlantique Nord* (La Géographie 40, 1923).
GREGORY, J. W. *The geological history of the Atlantic Ocean* (Quart. Journ. Geol. Soc. London 85, 1929).
GREGORY, J. W. *The geological history of the Pacific Ocean* (Quart. Journ. Geol. Soc. London 86, 1930).
GUTENBERG, B. and RICHTER, C. F. *Structure of the crust. Continents and oceans* (Ch. XII in Physics of the Earth VII, 1939).
HAUG, E. *Les géosynclinaux et les aires continentales* (Bull. Soc. Geol. de France, 3e sér. 28, 1900).
HÖGBOM, A. G. *Die Atlantislitteratur unserer Zeit* (Bull. of the Geol. Inst. of the University of Upsala, vol. 28, 1941).
JONES, O. TH. *Explorations in ocean basins* (Quart. Journ. Geol. Soc. 1937).
KOSSINNA, E. *Die Erdoberfläche* (Handb. der Geophysik II, 3, 1933).
LACROIX, A. *La constitution lithologique des îles volcaniques de la Polynésie australe* (Mém. Acad. Sci. France 59, 1927).
LACROIX, A. *Clipperton, îles de Pâques et Pitcairn* (Ann. de l'Institut Océanogr. N.Sér. 18, 1939).
LEAHY, L. TH. *The configuration of the bottom of the South Pacific* (C. R. Congr. Intern. de Géographie Amsterdam vol. 2, sect. IIA, 1938).
LITTLEHALES, G. W. *Configuration of the ocean basins* (Bull. Nat. Research Council 85, 1932).
MECKING, L. *Ozeanische Bodenformen und ihre Beziehungen zum Bau der Erde* (Peterm. Geogr. Mitt. 1940).
NANSEN, F. *Bathymetric map of the arctic basin* (Americ. Geogr. Soc. Spec. Public. 7, 1927).
NÖLKE, F. *Zur Tektonik des Atlantischen Beckens* (Geol. Rundschau 30, 1939).
SCHOTT, G. *Geographie des Atlantischen Ozeans* (1926).
SCHOTT, G. *Geographie des Indischen und Stillen Ozeans* (1935).
SONDER, R. A. *Zur Tektonik des Atlantischen Ozeans* (Geol. Rundschau 30, 1939).
SCHUCHERT, CH. *The problem of continental fracturing and diastrophism in Oceanica* (Americ. Journ of Sci. 42, 1916).
SCHUCHERT, CH. *Gondwana landbridges* (Bull. Geol. Soc. America 43, 1932).
SCHWINNER, R. *Lehrbuch der physikalischen Geologie* (I, 1936).
STILLE, H. *Zur Frage der transatlantischen Faltenverbindungen* (Sitzungsber. Preuss. Akad. Wissensch. 11, 1934).
STILLE, H. *Geotektonische Probleme im Atlantischen Raume* (Kais. Leopoldinische Carol. Deutsche Akad. d. Naturf. Halle 1937).
STILLE, H. *Queratlantische Faltenverbindungen* (Geol. Rundschau 30, 1939).
STILLE, H. *Kordillerisch-atlantische Wechselbeziehungen* (Geol. Rundschau 30, 1939).
STOCKS, TH. *Neues zur Morphometrie des Atlantischen Ozeans* (Ann. d. Hydrographie 67, 1939).
STOCKS, TH. und WÜST, G. *Die Tiefenverhältnisse des offenen Atlantischen Ozeans* (Meteor Exp. III, 1, 1935).
Symposium on the geophysical exploration of the ocean bottom (Proceed. Americ. Philos. Soc. 1938).
TERMIER, P. *Les Océans a travers les Ages* (Bull. Inst. Océanogr. Paris 1920); Idem in „*A la gloire de la Terre*" (1926).
THOM, W. T. *Position, extent and structural make up of Appalachia* (Bull. Geol. Soc. of America 48, 1937).
UMBGROVE, J. H. F. *Palaeogeographie der Oceanen* (Tijdschr. Kon. Nederl. Aardr. Genootschap 54, 1937).
VAUGHAN, T. W. *International aspects of Oceanography* (Nat. Acad. Sci. Washington 1937).
VENING MEINESZ, F. A. *Gravity over the Hawaian archipelago and over the Madeira area, conclusions about the Earth's crust* (Proc. Nederl. Acad. Wetensch. 44, 1941).
WEGENER, A. *The origin of continents and oceans* (New-York, Dutton 1924).
WILLIS, B. *Isthmian links* (Bull. Geol. Soc. America, vol. 43, 1932).
WISEMAN, J. D. H. and SEYMOUR SEWELL, R. B. *The floor of the Arabian Sea* (Geol. Magaz. 74, 1937).
WÜST, G. *Die Gliederung der Weltmeere* (Peterm. Geogr. Mitteil. 1936).

Chapter VII

ICE-AGES

"We are apt to think that our present conditions are normal, but that is far from being the case".

(A. P. Coleman).

Introduction

The earth's aspect is a very recent one from a geological point of view. Everyone knows that Greenland and Antarctica are at present covered by extensive ice-caps (fig. 71). Yet only one tick back on the geological clock large parts of America and Europe were covered by analogous ice-sheets of enormous dimensions. In Europe the ice-front extended from the British Isles to the Netherlands, and from there to Germany and North Russia, covering the whole of Scandinavia. Another ice-sheet covered the Great Lakes district of North-America and practically the whole of Canada (fig. 72). Indisputable traces of large glaciations in South-Africa, South-America, Australia and a few other regions date from a still more distant past — the Upper-Paleozoic (fig. 73), or even remoter times — the Precambrian, when plant-life had not yet imposed itself upon the land, and the seas were only populated by Invertebrates (fig. 74). Rocks whose origin cannot be explained unless we assume that they were formed under the same conditions of heat and drought as are now found in the Sahara (fig. 79) are met with in many places in Europe and America. On the other hand, the fossil remains of a luxuriant vegetation are still observed in such bitterly cold regions as Spitsbergen and Greenland.

The most absorbing topic of changing climates has not only attracted geologists, but also investigators in other branches of science, such as astronomers, meteorologists, paleontologists and biologists, and the hypotheses that attempt to find a solution for the earth's climatological problems are so numerous and so varied that Daly was induced to remark in this connection: "Geologists do not know the causes", and somewhat further: "At present the cause of excessive ice-making on the lands remains a baffling mystery".

However, the veil of mystery around these climates has been lifted to some extent by the increasing amount of data, and the minute studies and lucid suggestions of a large number of investigators. An attempt will now be made to unravel the main outlines of the chaotic mass of data and opinions and special attention will be paid to those which I consider to be the most

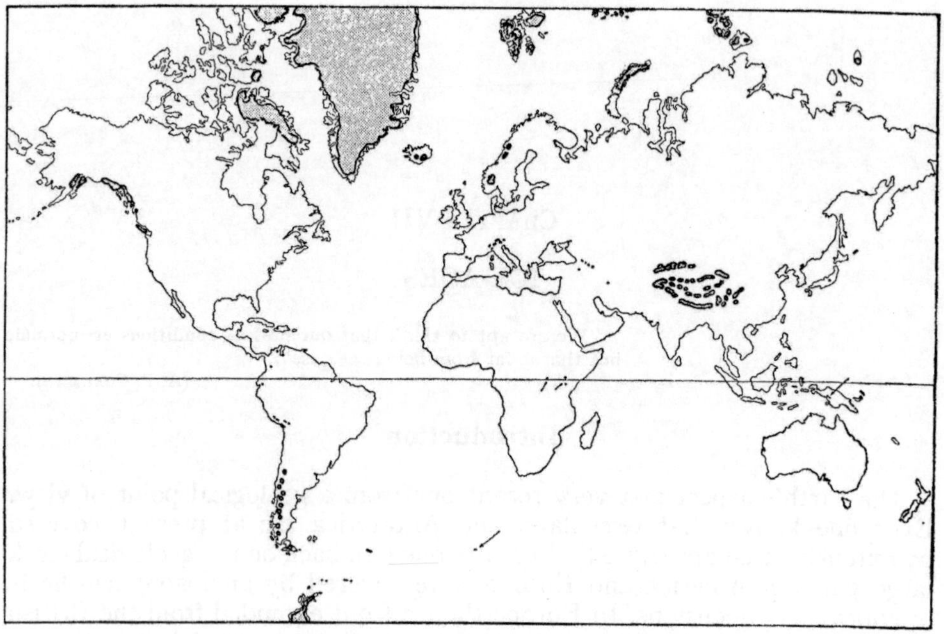

Fig. 71. Ice-sheets and glaciers of the present time.

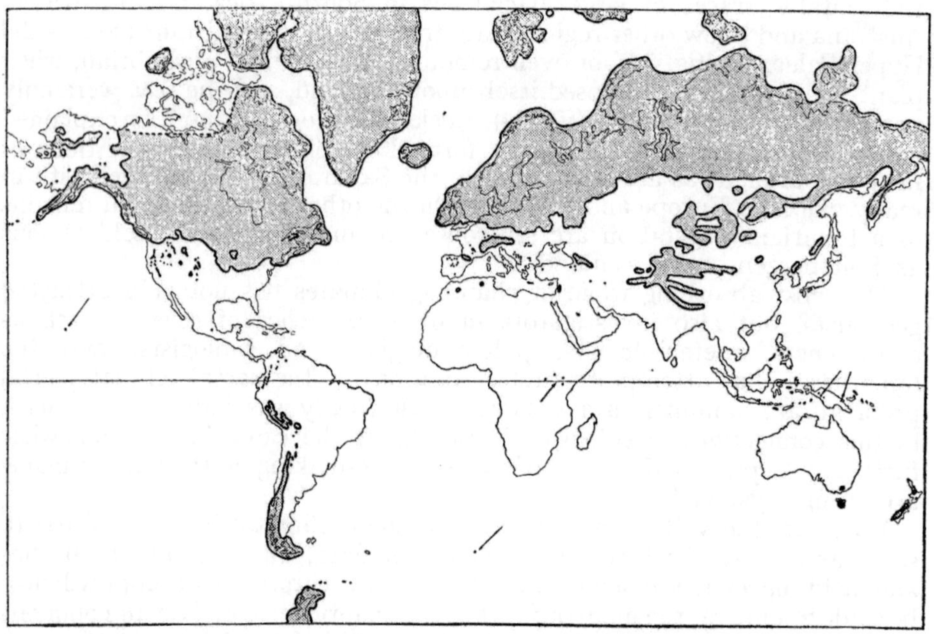

Fig. 72. Ice-sheets and glaciers during maximum glaciation in the Pleistocene.

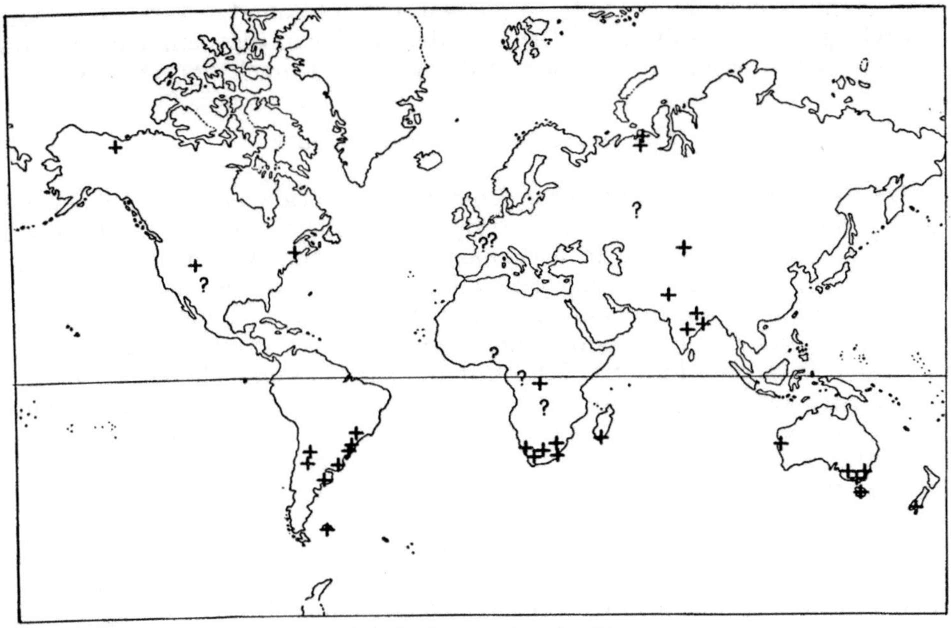

Fig. 73. Upper-paleozoic tillites.

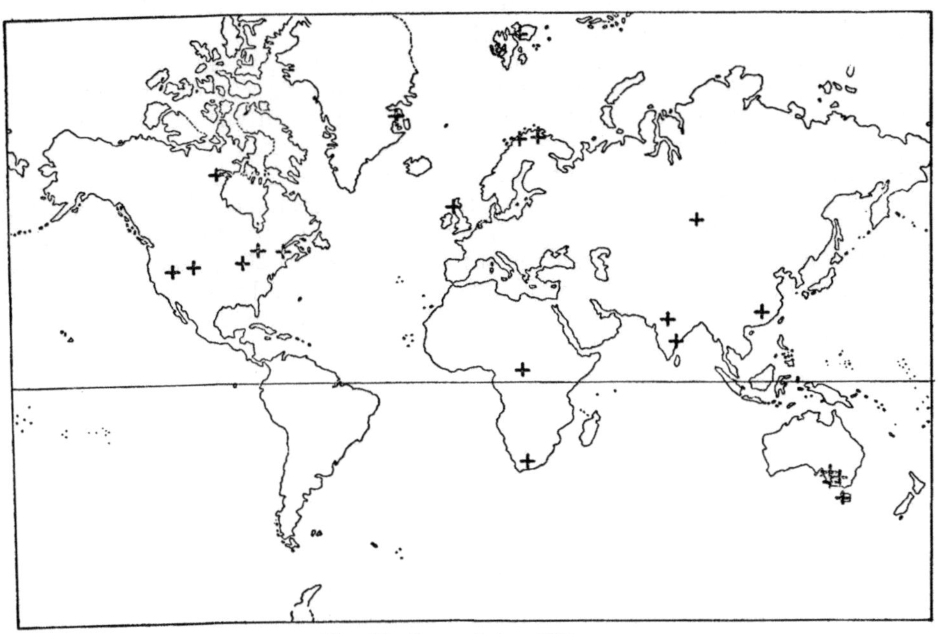

Fig. 74. Precambrian tillites.

important. I hope to be able to show the approximate state of present knowledge on the subject of the climates, and will also emphasize those aspects of the problem which are still wrapped in much uncertainty.

The periodicity of the climate

If noted in a proper order on the geological time-scale, periods of glaciation and periods that were characterized by a mild climate up to high latitudes will be seen to give rise to a curious phenomenon, viz. a large climatic periodicity. Some 250 million years elapsed between the beginning of the upper-paleozoic glaciation and the pleistocene ice-ages. Evidence of another extensive glaciation i.e. towards the close of the Precambrian, is found if we reach back an additional 250 million years. Holmes showed that this periodicity can probably be retraced several times in still remoter times, though there are as yet too few data to enable us to speak of a phenomenon recurring with perfect regularity [1]). A periodicity of 250 million years would appear to be a rough, but fairly accurate estimate.

The duration of a glaciation can only be indicated in one case, viz. in that of the Upper-Paleozoic, which continued for about 25 to 30 million years. Should the series of ice-ages, beginning with that of the Pleistocene (ca. one million years ago) be attributable to a phenomenon with a cause similar to that of the upper-paleozoic glaciation, the present period can consequently be said to constitute an intervening stage of partial deglaciation. Many glacial and interglacial stages may follow the present situation before a new period is reached with conditions comparable to those which existed from the Upper-Permian until the Pleistocene.

Speculation on what might happen if these ice-caps were to grow again excessively is one that cannot but fill one with dread. Prosperous agricultural and economic centers, thriving cities and important ore deposits would all disappear under the ice's irresistible march, and an alarming decrease in the areas fit for population would ensue. Yet let us suppose that the reverse were to happen, i.e. that the ice were to melt entirely. The resulting state of affairs would not be a much happier one. It can easily be calculated that the level of the sea would be raised some 40 to 50 [2]) meters if the ice were to melt over the whole world, and 12% of the land would consequently be flooded. Proud cities such as Amsterdam and Batavia would sink down in a muddy sea-bottom, and the lower parts of the sky-scrapers of New-York would be turned into shelters for lobsters. However, we are not concerned

---

1) The figures for the precambrian glaciations areapp rox. 500, approx. 600, approx. 800, approx. 1,000, probably > 1,000 and probably > 1,500 million years (A. Holmes, The age of the earth, 1937, p. 217).

2) If the ice were to melt over the whole world, the volume of water would amount to about 18 million cubic kilometers (Daly, p. 11), assuming a mean density of 0,9 for the land-ice. The total area of the oceans occupies 361 million square kilometers. This melting process would consequently cause the level of the ocean to rise 50 meters. The added weight of this water, however, would press down the ocean-floor to some extent, and the continents would be slightly elevated as a result of the displacement of subcrustal matter from the oceanic sectors. The change in the ocean-level would consequently ultimately amount to less than 50 meters.

with what might happen in the future, but with what happened in the past, and it is these last problems which we now intend to examine.

An intervening period between two galciations (e.g. from the Cambrian to the Upper-Carboniferous, and from the Upper-Permian to the Pleistocene) is marked by the absence of strongly differentiated climatic zones. The shape of the terrestrial globe has, of course, always been responsible for the fact that the polar regions constitute the colder areas of the earth, and the equatorial zones the warmer ones. The inclined rotation-axis of the earth has similarly always been responsible for more or less pronounced seasons. On the other hand, the seasonal changes during some periods were so small that several writers prefer to speak of a uniform climate for the whole world. The climate would appear to have been much milder up to a short distance from the poles in former times. This is shown (1) by the luxuriant tertiary vegetation in such high northern latitudes as Greenland, where the only existing plant life at present is the tundra, and (2) by the reef corals that flourished up to much more northern and southern latitudes. Though the latter are only found in warm seas, they are known to have existed as extensive coral reefs in the tertiary seas of North-America, and were distributed as far as Maastricht (and even Denmark) in the Senonian.

Special phenomena that occurred during certain periods show that the climate was less differentiated than during the major periods of glaciation. This is how some authors interpret the tillites, which may be observed in the eocene beds of North-America and Antarctica, the Jurassic of California (fig. 75), the Upper Silurian and Devonian of South-Africa and Alaska, the Ordovician of Europe and North America (fig. 76), and possibly a few other localities and periods (the Triassic of Central-Africa?) [1]). No one regards these tillites as the products of former ice-sheets. They are thought to constitute marks left by more or less local mountain glaciers and piedmont glaciers. It may still be questioned whether the formation of the glaciers in different regions dit not in each case result from a combination of local circumstances. There is reason to believe, however, that these old moraines of Eocene, Jurassic, Devonian, Upper-Silurian and Ordovician time justify the assumption that minor periods of cold prevailed during the long lapse of time that separated two major periods of glaciation. This enables us to draw a *schematic* curve of the climate such as that in Table II.

A mild and equable climate, affecting almost the whole world and extending even as far as polar latitudes, prevailed during such aeons of our earth's history that this condition may be considered to have represented the earth's normal climate. The major glaciations, on the other hand, and the minor periods of cold which we have just mentioned may be assumed to form "abnormal" and relatively short deviations.

The earth's normal climate

What was the earth's aspect during normal climatic conditions? The remains of a mesozoic vegetation are found in Greenland, and as far as

[1]) Fig. 71–76 have been adapted from publications by Antevs, Backlund, Brooks, Coleman, Norin, Teichert and others.

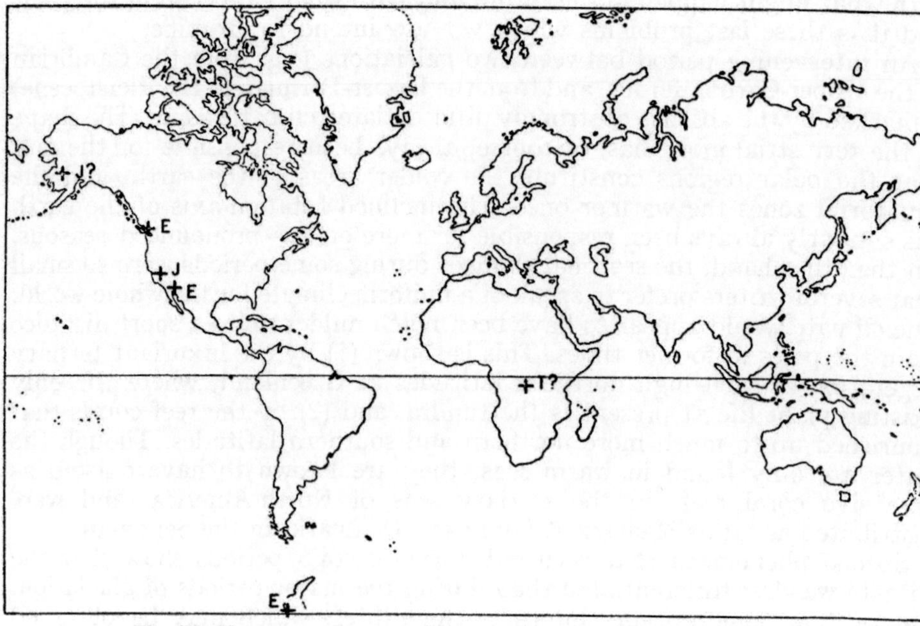

Fig. 75. Mesozoic and eocene tillites. T Triassic?, J Jurassic, E Eocene.

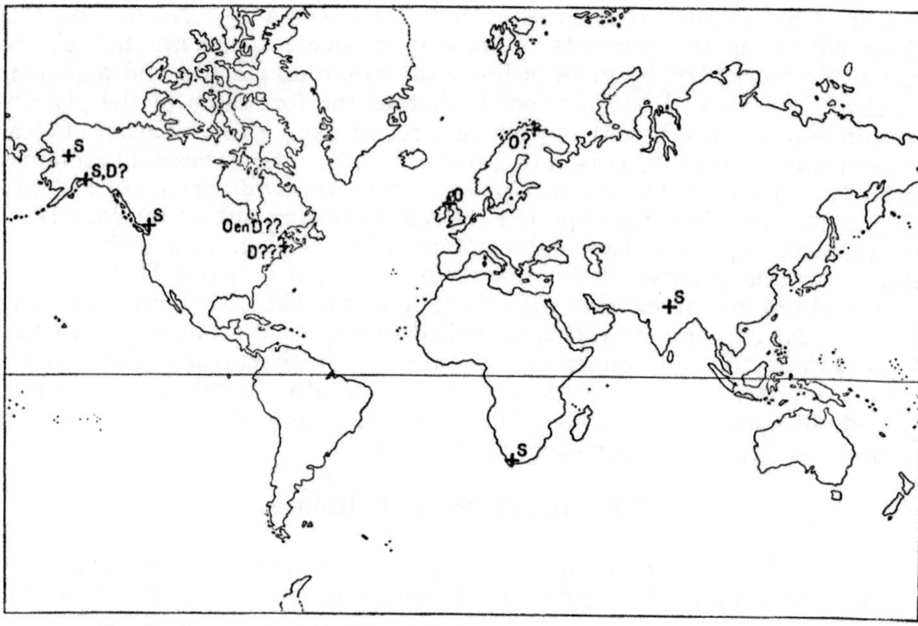

Fig. 76. Lower-paleozoic tillites. O Ordovician, S Silurian, D Upper Devonian.

Grahamland, also between (and including) New-Zealand and North-America. Darrah [1]) elucidates this question as follows: "The Jurassic floras of such widely separated regions as the Antarctic continent, England, India and North-America have so very many plants in common that the essential uniformity of the vegetation all over the earth at that period is well known. The succeeding flora of the Cretaceous displays almost as great a uniformity in the composition. Not only were these floras homogeneous, but they appear to have flourished under more equable conditions than those now prevailing over most of the earth." A few pages further, Darrah states: "The rapid radiation of the diversifying flowering plants (of the Upper-Cretaceous) certainly is indicative of an equable climate and an absence of barriers of migration."

Recent discoveries have made it clear that the climate need only change a little to bring about a recurrence of a luxuriant growth up to such high northern latitudes as Greenland. Since the retreat of the ice towards the beginning of this century, the remains of an autochthonous vegetation and of settlements dating from the eleventh and twelfth centuries have been brought to light in this region, and indicate that at that time the transitory melting of the ice-caps took place on a much larger scale than during the last forty years.

Under the mild climatic conditions of the Mesozoic, the land-fauna, including organisms such as the giant reptiles, were able to migrate from Central-Asia to America (via the Straits of Bering) and Africa without meeting any unusual barriers of temperature (fig. 77).

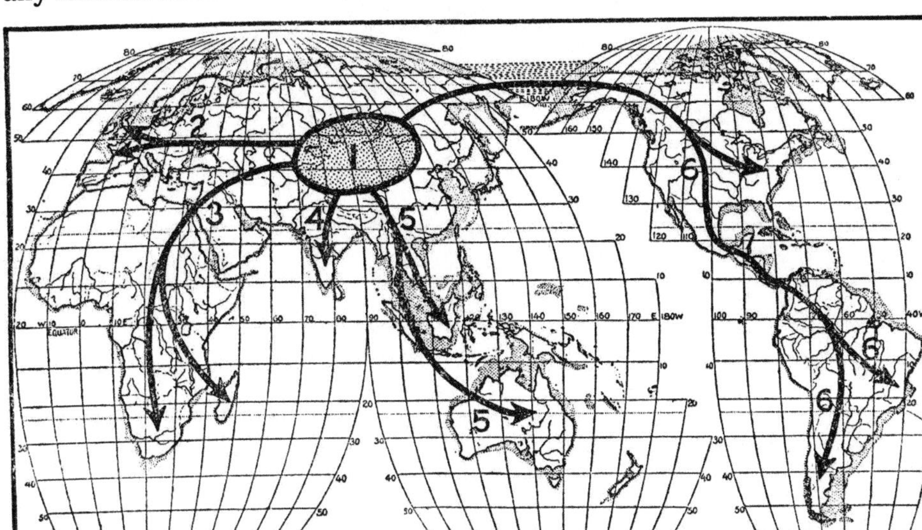

Fig. 77. Upper-jurassic — lower-cretaceous routes of migration of the giant Dinosaurs (After H. F. Osborn).

Such an evenly distributed climate also influences the atmospheric and

1) Darrah, 1939, p. 171, 180.

oceanic circulation, producing less intensive horizontal and vertical movements. One of its consequences was that the deep-sea received a smaller supply of oxygen. It may be assumed that marine animals that had adapted themselves to colder water could only have thrived in the northern and southern polar seas. These have indeed been described from jurassic and lower-cretaceous deposits, and are known as the boreal fauna. A specific geographical position may cause this boreal fauna to extend locally up to a shorter distance from the equator than elsewhere. The remaining marine fauna is almost entirely uniform. Everyone will agree that though Neumayer distinguishes between four types of Mesozoic marine faunas: the Mediterrane Caucasian, Himalayan, Japanese and South-Andine "provinces", his classification merely divides the biocoenoses of the warm seas into separate geographical provinces. The cambrian Trilobites, too, for instance, could be subdivided into an Atlantic and Pacific province, but this does not mean that they lived under different climatic conditions.

The reef-building Rudists of the Upper-Cretaceous (fig. 78) have often been

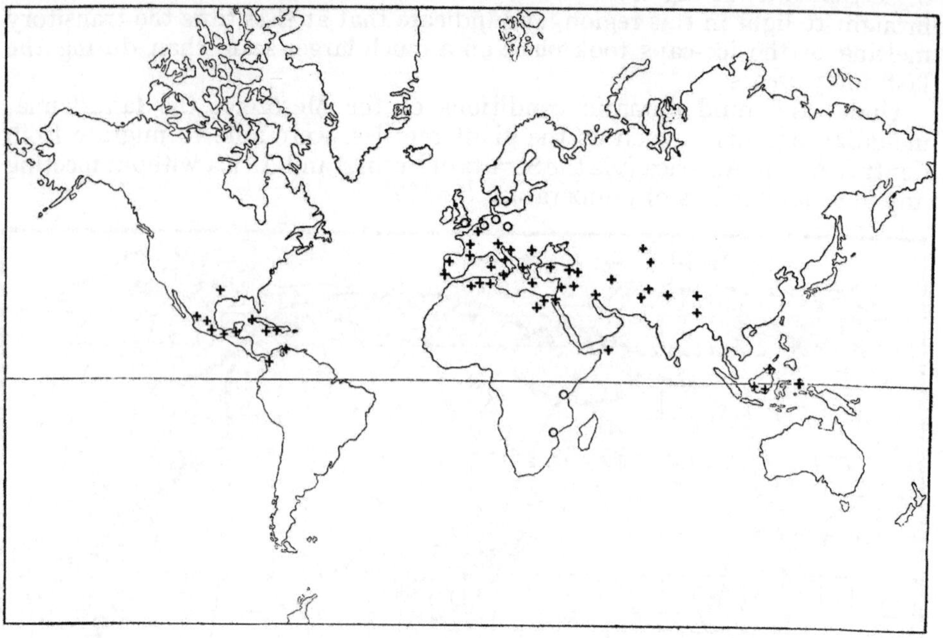

Fig. 78. The distribution of upper-cretaceous Rudists, + normal types, O dwarf types (After Dacqué).

cited as examples of warm-water organisms. It is noteworthy that dwarf-forms have been found in addition to the larger normal types in their most northern occurrences. Some attribute this to the influence of the boreal zone. A like phenomenon is observed in their southern extension, in East-Africa, where they occur from lat. 5° to 20° S (lat. 50° to 55° N. on the other side of the equator). It is clear that the existence of a southern cold sea-current coming from Antarctic regions must be taken into account if their distribution is to be explained.

It would not surprise us if those zones of the earth which may be said to be predestined to become belts of arid degradation had had a greater extent during a "warm period" than they have at present. There can hardly be anything unusual in the distribution of the deserts during a period of normal geological climates if we remember that the present distribution of land and sea still exerts a strong influence on the disposition of arid areas. Fig. 79 indicates the location of former mesozoic desert formations, together

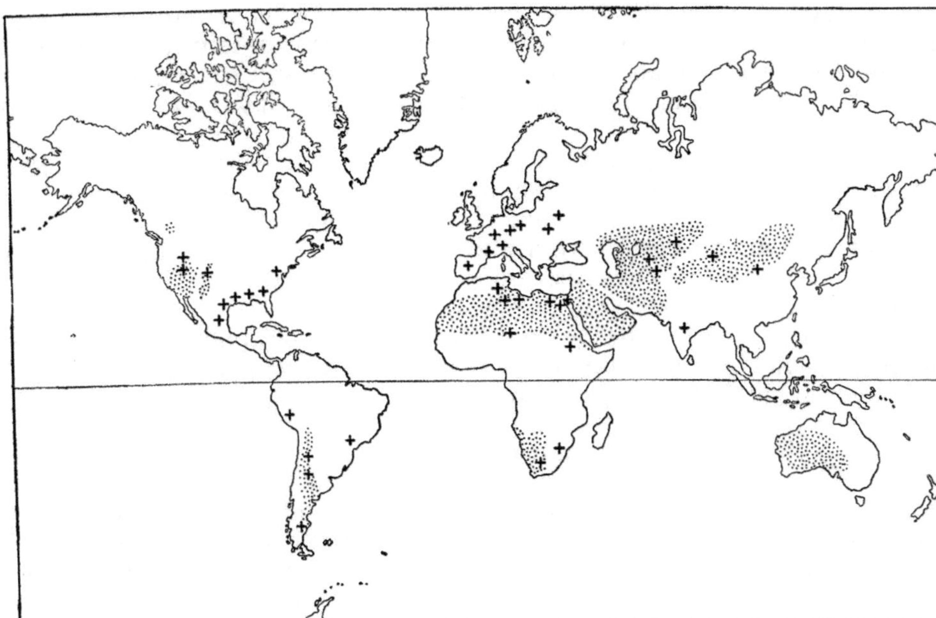

Fig. 79. Desert deposits of Mesozoic times (+), after A. Wegener, and desert belts of the present time (stippled).

with those areas which are now occupied by deserts. Brooks pointed out at great length that the earth's normal climate need not be the result of an important change in the position of the continents and poles. Berry [1]) illustrated this statement as follows as regards the distribution of the flora: "Fossil plants have been found at six general horizons in the Arctic, namely: late Devonian, Mississippian, late Triassic, late Jurassic and early Lower Cretaceous, Upper Cretaceous, and Eo-Oligocene. They are frequently associated with coal seams, and this, together with roots in place in the underlying clays, the presence of fresh-water diatoms, mollusks and aquatic beetles, proves that the plants grew near where they are found fossil and were not drifted from lower latitudes". .... "The distribution of the localities

1) Berry, 1929, p. 236.

around the poles conclusively discredits the hypothesis, now fashionable in certain quarters, that there has been a change in position of the pole. A botanical analysis of the floras shows clearly that the term tropical, so frequently applied to them, is highly improper, and that they are, for the most part, distinctly cold temperate. The associated petrified woods, with the single exception of the *Dadoxylon* from the Mississippian of Spitsbergen, show well marked seasonal rings.

"Detailed comparisons of these Arctic floras with contemporaneous floras from lower latitudes, especially in the case of the younger, post-Paleozoic floras, show unmistakable evidence for the existence of climatic zones. It was further shown that these northward extensions of temperate floras in the past were all contemporaneous with or immediately followed times of unusually widespread continental submergence and sea extension, and that they lived under the domain of oceanic as opposed to continental climatic conditions.

"It was concluded that low lands and expanded seas were amply sufficient to account for the necessary ten to fifteen degrees northward swing of isotherms demanded by the botanical character of the floras. This, with the concomitant alterations in wind and current circulation, combined with Brooks' meteorological results, are considered to account for all of the observed facts."

Nor need we revert to the idea of continental drift to explain the occurrence of periods of abnormal climate or glaciations, to which we will confine ourselves in the following paragraphs.

The abnormal character of present conditions

It should be remembered that we are living in an abnormal climatological age, albeit in an intervening stage of milder glaciation or partial deglaciation. The actual distribution of fauna and flora in distinct biogeographical units determined by the climatic zones is a phenomenon of very recent origin, and merely dates from the close of the Tertiary (cf. Chapter VIII). The extreme specialization and differentiation of the flora and fauna forms another unusual aspect of our time (cf. Chapter VIII). For example, flowering plants (Angiospermae), which were first observed in the Upper-Mesozoic, dominate the plant-world, while the adaptive radiation of the mammalian fauna only began to develop in the Tertiary. What do we know of the ecological and biological conditions necessary for the maintenance of a biocoenosis during the Mesozoic and Paleozoic? It has frequently been shown that an attempt to base conclusions as regards the climatological significance of fossil finds on present conditions is bound to meet with failure. The rhinoceros, and a type of elephant — the mammoth — populated north-western Europe during the bitterly cold Pleistocene. Yet these animals' near relatives are now found exclusively in the tropics! This in itself is quite enough to show that the greatest care should be taken by investigators before basing conclusions on the present situation. It was shown above that a very different atmospheric and oceanic circulation must have prevailed formerly, during periods of a normal climate. It will also be obvious that far more intensive denudation must have influenced the land during a period like the Lower-Paleozoic, for no vegetation protected the earth at that time. These examples may help one to obtain some idea of the many and occasionally very difficult problems which confront the paleontologist and climatologist in their attempts to interpret the geological past.

This point needs emphasizing when an attempt is made to find a solution for the climatological changes in the earth's history. For many authors put the problem in such a way that two wholly different questions have to be solved simultaneously. The first deals with the cold ages, the second with the warm climates. These scientists begin by using the present climatic conditions as a basis upon which to build further conclusions, and then try to explain both the glaciations and the cause of the climatic change which transformed the climate into an almost mild and equable one all over the world. The problem should be put quite differently, however, for in following the above method we assume the present situation to be the normal one and to characterize the whole geological past, and thus take it for granted that any other situation must be an abnormal one. Investigators such as Köppen and Wegener undoubtedly lost sight of the abnormal character of actual conditions when determining the climatic zones of former geological periods, for they sought to reconstruct climatic zones equivalent to our own for all these formations, and any conflicting evidence in the past was simply attributed to hypothetical changes in the position of the poles and continents. They thus construed a continuous, non-periodical process, instead of the periodicity of abnormal climates.

The problem

As already stated, the whole question should be approached from a different angle. We should break with the habit of considering our present conditions to be normal ones. Once this is done, the main problem left is: what caused the abnormal or cold deviations of the earth's climatological history? In other words, to what impulses should we attribute the glaciations? The problem is fourfold, for the following points have to be explained: (1) the major periodicity of approximately 250 million years, including the intervening minor cold spells; (2) the relatively rapid growth of an ice-cap and its subsequent equally rapid decay; (3) the occurrence of interglacial stages; and (4) the site of an ice-cap.

The relation between periodicity and physico-geographical factors

Many authors have tried to connect climatic changes with other events within the earth itself, or with processes upon its surface. The pleistocene ice-ages succeeded the folding of such belts as those extending between the Alps and the Himalayas. The upper-paleozoic glaciations appeared as the Variscian Mountains began to emerge, and the glacial phenomena of the Precambrian coincide with the so-called Huronian period of mountain-building. The level of the continents was a comparatively high one compared to that of the sea. Glaciations can thus be said to be observed during periods in which we find much land and few epicontinental seas. The intervening ages, however, are characterized by a relative tectonic quiet and low continents, large parts of which were flooded by the seas. These periods were marked by a mild climate up to high latitudes. The relation of epochs of mountain-building to glaciations was revealed by Leconte in 1899, and in the same way Wright, Upham, and especially Ramsay and Ruedemann attempted to explain the phenomenon of glaciation. These authors assert that the ice-caps originated as a result of the extension and particularly the altitude of land masses. Before going any further, however, a few other phenomena, which are likewise related to periods of mountain-building and oscillations of the sea-level, will have to be mentioned, since these enable us not only to form a particularly clear picture of the peculiar intricacies of the problem, but also of the various lines along which a solution has so far been sought for the occurrence of glaciations.

Several writers stress the possible influence of a changing amount of carbon-dioxide in the atmosphere. This gas, which is transparent to light, though opaque to heat, would act in the atmosphere like a glass pane in a hot-house and retain the heat, and the temperature would then increase proportionally to the quantity of CO_2 present in the air (Arrhenius). Chamberlin claims that a considerable amount of CO_2 was locked up in the formation of limestones and dolomites during periods of continental emersion (regression), and that the climate was consequently transformed into a cold one. On the other hand, periods of submersion, or peneplaination and transgression would have set this CO_2 free, and this would then have produced a milder climate.

The course of the sea-currents has undoubtedly changed (and the salinity of the oceans oscillated to some extent) in the course of time. Changes have doubtlessly also affected the atmospheric circulation and the cloudiness of the sky. All these variations were due to the repeated alterations in the distribution of land and sea.

Other investigators (especially Humphreys) argue that a great quantity of fine dust is flung into the atmosphere — and even the stratosphere — during violent volcanic eruptions, and that the dust continues to linger in these higher regions during a considerable lapse of time, scattering and reflecting the solar radiation. Since these eruptions are known to have occurred with exceptional frequency during certain periods, it follows that the atmosphere may have been cooled during these stages.

Many factors had an undeniable influence on the climates during the changes which were constantly affecting the face of the earth. It is almost impossible, however, to describe each one's influence separately, or to estimate their combined total effect quantitatively, one of the chief difficulties being that our paleogeographical maps are incomplete in several respects, and also very schematic. Yet it cannot be denied that mountain-building and the major periods of glaciation are in fact correlated in one way or another.

Still, intensive folding and subsequent mountain-building are known to have occurred during several other periods — e.g. the Caledonian, the early and late Cimmerian (Nevadian), and the Laramide. As stated above, a few of the less important glacial deposits may possibly be related to these periods. The tillites of Ordovician time and the Silurian and Devonian have been assumed to be related as such to caledonian mountain-building, and those of the Jurassic and Eocene to the late cimmerian and the laramide revolution respectively. The Variscian and Alpine Mountains (as shown in Chapter II) were far more extensive than the Caledonian, Cimmerian and Laramide chains. Though it is extremely difficult — and at times almost impossible — to reconstruct a satisfactory and comprehensive map of the above periods, the following conclusions may safely be drawn: (1) that the periodicity of 250 million years as indicated by the major periods of glaciation is also manifested in the epochs of mountain-building (2) that a rhythm of minor amplitude, as observed in the less accentuated deviations from normal climatological conditions in the Ordovician, Silurian, Devonian (Triassic?), Jurassic and Eocene can probably be said to coincide with epochs of mountain-building of far lesser geographical importance than those which accompanied the major rhythm of 250 million years referred to above. This is illustrated by the schematic graph in Table II [1]).

The deeper causes of periodicity

A comparison between the various processes of folding and movements of the sea-level shows that these heterogeneous phenomena are distinctly correlated. There also appears to be some relation to magmatic cycles; and the alternating increase and decrease of compression of the earth's crust seems to be directly connected with the above events, as explained in the preceding chapters. In short, everything shows that a whole series of different phenomena reflect the influence of a deeper and wide-spread impulse. The term "deeper" should be taken in its figurative and literal sense, for the phenomena are attributed to deep-seated forces. It would take us too far to examine this difficult and intricate problem at this point, but it should be noted that several diverging hypotheses have been put forward on this subject by Joly, Holmes, Bucher and — recently — Griggs, which though differing in many respects, agree on one point, for each attributes the origin of the phenomena to subcrustal forces. This vague designation will have to suffice for the moment. However, this much can be said now: the periodical processes in the earth's interior may be thermal cycles, i.e. they may be accompanied by periods of an increased accumulation of heat alternating with periods of decreasing temperature. Folding would then have occurred in the last phase. This hypothesis implies that it cannot be a mere accident that these phenomena preceded periods of mountain-building and the abnormal "cold" climates.

It should not be assumed, however, that a fall in the temperature of the atmosphere is due exclusively to these subcrustal conditions. For there still remain the changes in the earth's physiographical aspect (mountain-building, regression, a decrease in the amount of CO_2 in the atmosphere, a less intensive

---

1) Similar illustrations are found in Schuchert, Dacqué and Brooks.

circulation, etc., all of which have a considerable influence on atmospheric conditions). To these should be added a further factor, which is known to intensify the fall of the annual mean temperature once the latter has begun to decrease. This factor will now be examined.

The rapid growth and subsequent retreat of an ice-cap

Brooks argues that the ice-cap has a cooling effect on the atmosphere above and along its edge and that this will become an important factor once the ice-cap has begun to grow. The additional cooling influence will cause the ice-cap to extend beyond the boundaries which would have existed without such an effect, i.e. it will cause it to expand beyond the polar regions where an initial fall in the temperature would result in the formation of an ice-cap. On the other hand, the temperature is known to rise steadily in the areas between the poles and the equator, and a state of equilibrium will consequently be reached at a given moment between both counteracting factors. An exhaustive review of this subject will be found in Brooks, but I would like to draw attention to a few of this author's results.

The ice-cap extends at an increasing rate once it acquires a given size, for its area increases by the square of its radius. The opposite effect, however, — i.e. the rising temperature from the poles to the equator — is a gradual one and proportional to the distance from the pole (normal horizontal gradient). Observations in polar regions seem to bear out this statement, revealing that the cooling effect of an extending ice-cap will in fact increase almost proportionately to its surface till the growing circumference of the ice-cap has attained a radius of approximately ten degrees. The more distant regions cease to exert their full effect in case of a larger area. For though the cooling effect will continue to increase, the average effect — expressed in km² — will begin to diminish (see the following table and fig. 80).

Radius of ice-cap, in °	Total cooling on edge of ice-cap, °F	Rise of temperature due to normal horizontal gradient, °F
0	0,6	0,0
1	1,05	0,9
2	2,44	1,8
3	4,6	2,7
4	7,8	3,6
5	11,8	4,5
6	14,5	5,4
8	17,2	7,2
10	18,2	9,0
15	20,4	13,5
20	21,3	18.0
25	22,2	22,5

Brooks concludes that if the initial winter cooling of an open polar ocean drops to a mere 0.6° F. (0.35° C) below the water's freezing point, the result will be the formation of an ice-cap, which at first extends rapidly to a latitude of approximately 78° and then more slowly to about 65° (i.e.

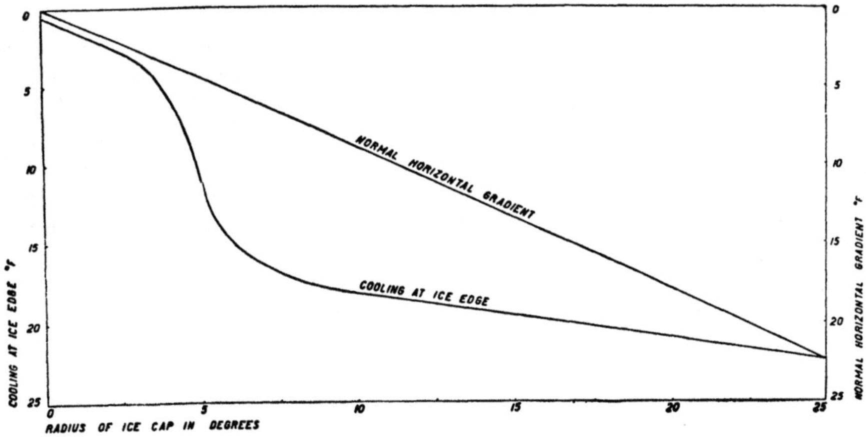

Fig. 80. Cooling power of an ice-cap (After Brooks).

assuming that no land influences or warm sea currents arrest its growth), and the ultimate lowering of the winter temperature, owing to the initial fall of 0.6° F., will then amount to 45° F. (25° C.). This applies to ice-sheets on sea as well as to those on land. Brooks attached such importance to the cooling power of an ice-cap that he devoted the first chapter of his book to this subject.

This much seems clear, however: the annual mean temperature will, once it has begun to fall as a result of some cause, be lowered still further the moment an ice-sheet is created.

The same so-called cooling power of an ice-cap may probably also explain the extremely rapid growth and subsequent equally swift decay of the pleistocene ice-sheets. The retreat of the last Scandinavian ice-cap has been depicted in fig. 81 and 82. The effect in this case is seen to consist of a slow and subsequently faster retreat of the border of the ice-cap. It should be noted, however, that another factor helped to accelerate the retreating movement. For the ice-front began to yield at an increasing rate as the ice-cap melted and grew thinner. The extreme right of the curve in fig. 82 finds its answer in one more factor. The land began to rise as it was freed of the load of the ice-sheet, and the highly elevated remnants of the ice-cap consequently melted at a much slower rate.

The following constitutes another important point, in my opinion. An extensive retreat of the sea is one of many factors that initiate glaciation. Once the ice-caps begin to extend, however, they will cause the sea-level to drop. The maximum lowering of the sea-level by glacial control must have attained approximately 100 meters during the Pleistocene. We know now that a great many shallow seas were reclaimed by the continents during

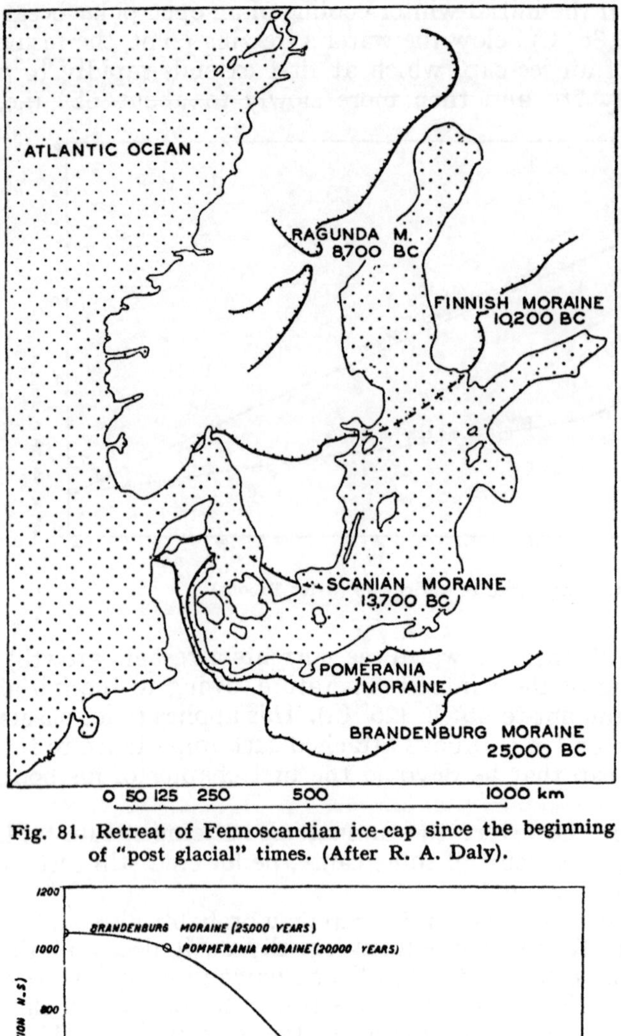

Fig. 81. Retreat of Fennoscandian ice-cap since the beginning of "post glacial" times. (After R. A. Daly).

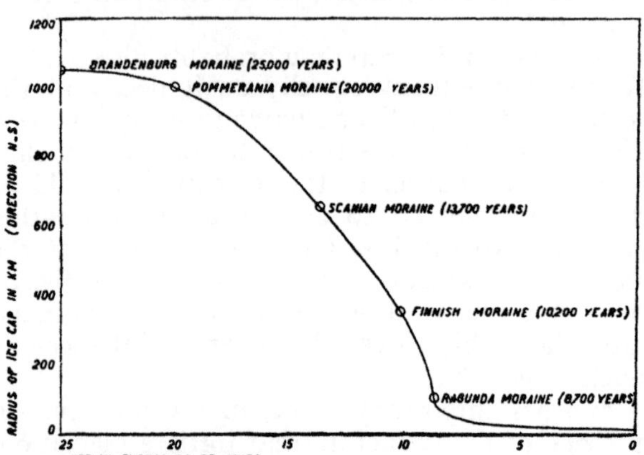

Fig. 82. Graph of the retreat of the Fennoscandian ice-cap, based on fig. 81.

this period, and that these same areas were drowned once more in post glacial times as the ice-caps began to melt. An example of this is found in the drowned rivers of pleistocene Sunda-land (fig. 83). This extension of land-areas as a result of glacial lowering of the sea-level favours glaciation.

Interglacial periods

The third aspect of the problem—viz. the interglacial periods that divide one long period of glaciation into a series of ice-ages—will now have to be examined. It is possible to distinguish between four separate ice-ages, — including three interglacial stages and our own "post-glacial" period (this may appear to be an euphemism!) — from the Pleistocene up to the present day.

The investigations by Milankovitch and Spitaler are of particular importance for this part of the climatic problem. Milankovitch calculated the radiation of the sun during the last 600,000 years. The heat which different zones of the globe obtain from the sun depends on three astronomical variables — viz. the obliquity of the ecliptic, the eccentricity of the orbit of the earth, and the cyclic variation of the perihelion. The obliquity of the ecliptic is responsible for the seasons. The greater the obliquity, the more contrast we find in

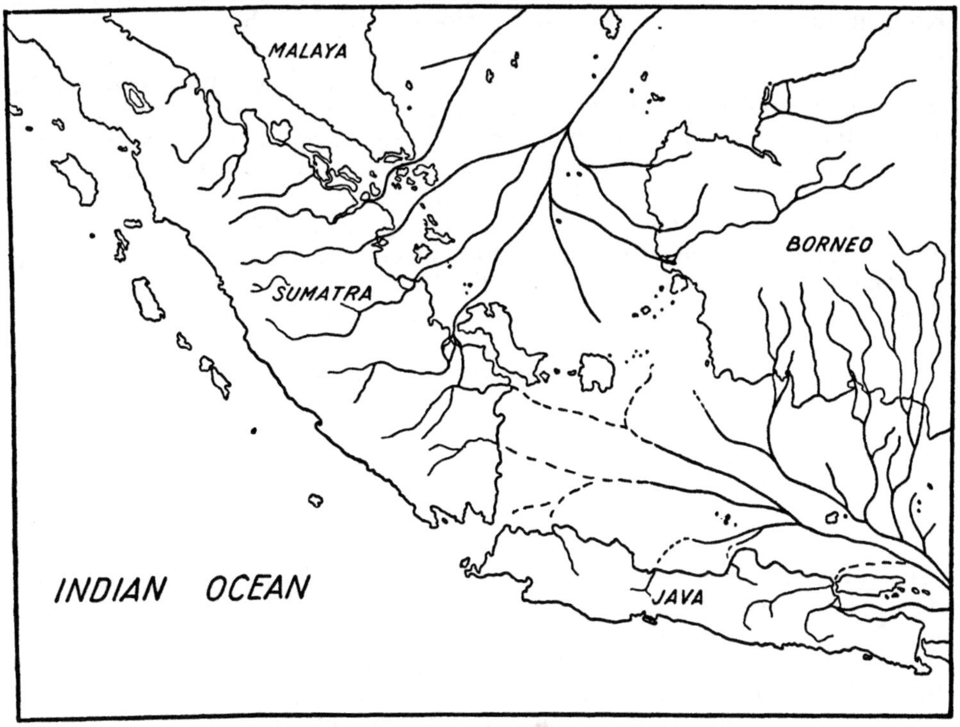

Fig. 83. Drowned rivers in the Java Sea and South China Sea.

the seasons. The limits of variation vary between 22° and 24$^1/_2$° in 40,400 years according to Milankovitch.

The orbit described by the earth around the sun is an ellipse, with the sun at one of the foci. The distance between the focus and the center of the ellipse is called the earth's eccentricity and, expressed in terms of the major axis, varies between zero and approximately 0,07 in about 100,000 years .The third variable is the cyclic variation of the perihelion with a period of 20,700 years, i.e. the so-called precession ot the equinoxes. Should the maximum effects of these three variables coincide the result would be a minimum radiation. It was on this basis that Milankovitch calculated and constructed his radiation-curve for the sun. Said curve contains a quantity of small peaks and several higher ones. Some authors — particularly Köppen and Soergel — claim that a remarkable similarity exists between this curve and the conclusions regarding glacial and interglacial stages arrived at previously by geologists.

Nevertheless, it should not be concluded that the problem can be considered to be definitely solved, for several circumstances will dampen our enthusiasm in this respects. To begin with, the curve is far from simple, and can be interpreted in several ways. It makes no reference to the way in which the combined atmospheric factors (such as circulation, cloudiness, evaporation and absorption) react to a greater or lesser amount of radiation. Köppen, for instance, supposes that periods with a less than normal summer temperature and short, mild winters favour the development of ice-sheets, but Croll advocated the exact opposite — i.e. that periods with exceptionally cold winters and short, warm summers provide the most favourable conditions for the evolution of such sheets. Besides, Spitaler's curve deviates to such an extent from that of Milankovitch's, that both may be said to agree only superficially. These authors thus arrive at very different results as regards the chronology of the ice-ages. The following table shows the astronomical time-scales of the four glacial stages as reconstructed by Köppen and Spitaler, and includes a list based on geological evidence, after Brooks. The figures are in thousands of years B.C.

	Spitaler	Köppen	Brooks
Fourth Ice-Age	250–150	188– 66	50– 28
Third Ice-Age	600–450	236–183	140–110
Second Ice-Age	1,000–800	478–429	430–380
First Ice-Age	1,350–1,150	592–543	—

The problem of radiation was likewise examined by Brooks. The relation of solar radiation to the appearance of sunspots is undoubtedly a complicated one. Moreover, it is difficult to predict the effect which a change in the radiation of the sun will produce on a climate. Notwithstanding these difficulties, Brooks may safely be said to have shown a distinct connection between solar variation and general atmospheric changes. A great sunspot maximum marked the year 1870, and since then (apart from certain fluctuations, such as an eleven-year cycle) there has been a gradual decrease of sunspots. During this time the average annual temperature and pressure rose in North-Asia, Europe, Iceland and part of North-America. Moreover, the glaciers and ice-caps have, on the whole, been retreating — especially since the beginning of the twentieth century.

In fine, all that can be said at present is that glacial and interglacial stages may possibly, and even probably, be attributed to changes in the amount of radiation dependent upon several cosmic variables, and that a correct interpretation of the climatic effect of these variable factors, and an accurate reconstruction and interpretation of the radiation-curve, must remain reserved for the future.

Of course, these astronomical factors were already active before the beginning of the Pleistocene. Yet, the fact that they only produced ice-ages during certain distinct periods proves that glaciation must have been initiated by one or more additional causes. The latter were discussed above, and it was pointed out that the major periods of abnormal climates are controlled by geographical changes and deep-seated terrestrial processes.

The geographical location of an ice-cap

As mentioned above, the last part of the problem deals with the site of an ice-cap once glaciation has become possible.

A few examples were given of normal climates, and it was seen that arid regions, for instance, will change with corresponding alterations in the distribution of land and sea.

If the earth's surface were homogeneous, with the exception of an ice-cap stationed in the present areas of Greenland and the South Pole, the atmospheric circulation would be approximately equivalent to that shown by Hobbs in the accompanying diagram (fig. 84). The actual situation deviates from this theoretical conception because of the special distribution of land and sea.

In this schematic diagram two zones of high pressure will be seen to run parallel to the equator, and these are predestined to become belts where the climate is subjected to an arid degradation. The present geographical conditions, however, are responsible for the fact that this is only partly true. To mention two examples: they caused the arid zone to shift northwards in Central Asia, and the same conditions were responsible for the absence of an ice-cap in Siberia, though this last area is known to constitute the coldest region on earth. The distribution not only of the temperature, but also of pressure, water vapor and the whole complex of oceanic and atmospheric circulation has a most important bearing on these deviations from the schematic diagram.

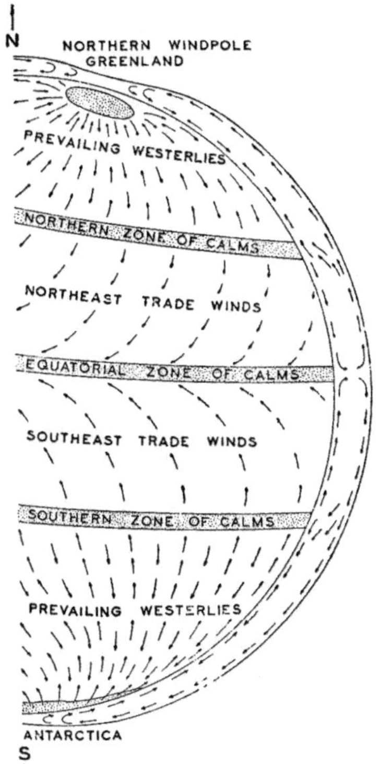

Fig. 84. Diagram of the atmospheric circulation (After W. H. Hobbs).

This will also help to make it clear why the large pleistocene ice-sheets are found in the northern hemisphere, and Brooks even tried to show why they originated in those localities in which they are actually known to have occurred. To understand his arguments, it will be necessary to return once more to the influence of solar radiation. As mentioned above, Brooks revealed that a decrease in the amount of sunspots since 1870 has clearly been accompanied by a corresponding increase in the mean annual temperature and pressure over specific areas in North-America, Iceland, Europe and Northern Asia. With the aid of observations obtained from these regions, and the logical premiss that a growing number of sunspots would have a contrary effect, Brooks constructed two maps (fig. 85 and 86), showing the distribution of the temperature and pressure in accordance with a

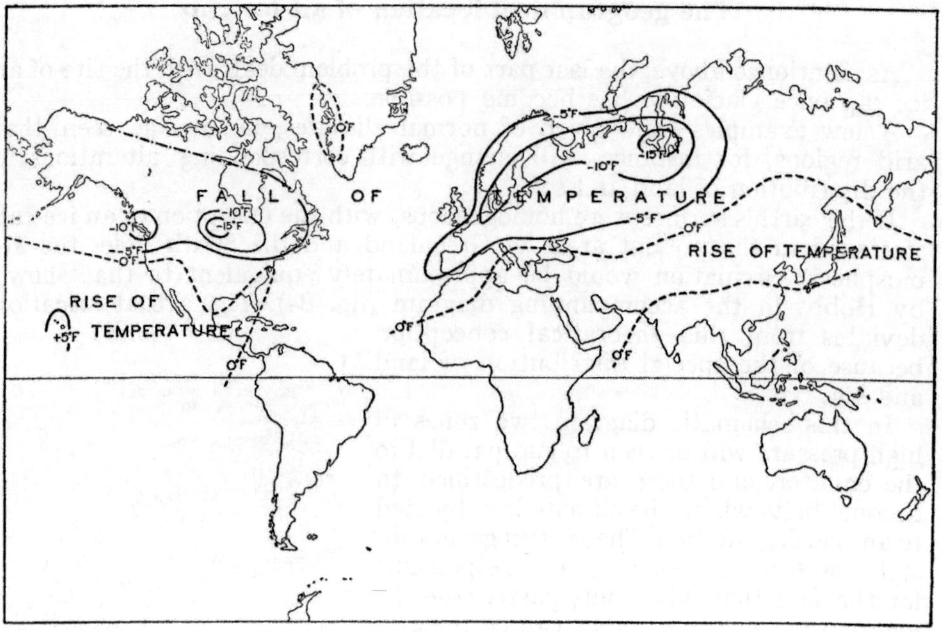

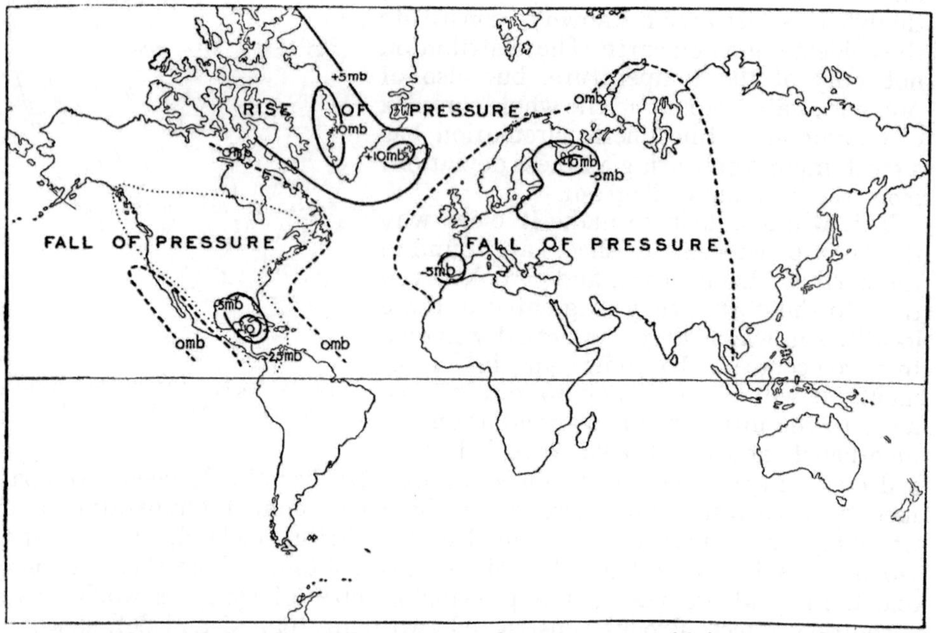

Fig. 85 and 86. Changes of mean temperature and mean pressure with increase of the relative sunspot number to 200 (After C. E. P. Brooks) mb = 1 millibar = 1,0197 g/cm².

hypothetical increase in spottedness up to an average of 200. This "relative sunspot number" means that approximately one five-hundredth part of the sun's visible disc is covered by spots. The figure for the sunspot maximum in 1870 was 139, and 32 for the eleven-year period from 1904 to 1914. Fig. 85 indicates a fall in temperature of more than 5° F. (approx. 3° C.), and of even as much as 15° F. (approx. 8° C.) over Canada, Northern Europe and the western part of Siberia, together with a rise in the temperature above the regions situated south of these areas. Fig. 86 depicts a decrease in the pressure over the continents, and an area of increased pressure over the oceans. These reconstructions present conditions which may very well have prevailed towards the beginning of the pleistocene ice ages!

It should not be forgotten that these results are based on the present distribution of land, sea and mountain belts. There is indeed good reason to suppose that the position of the continents during the Pleistocene did not differ from their present situation. However, let us reach back somewhat further into the past. The plant-bearing deposits of Eocene time causes Chaney to conclude that the position of the European continent in respect to America has remained unaltered since as far back as the Eocene. The accompanying maps make it easy to follow his argumentation. Those eocene plant units that bear evidence of the formerly existing climates have been grouped in fig. 87, and "similar fossil floras, which are considered to represent essentially similar climatic conditions" have been connected by lines which Chaney referred to as isoflors. These are also felt to represent eocene isotherms. Fig. 87 should be compared with fig. 88, which includes present-day isotherms for the month of January and the modern flora-types. The similarity between our isotherms and those of Eocene time is apparent, and Chaney consequently wrote: "Present departures of isotherms from coincidence with parallels of latitude are interpreted from the tempering effects of ocean currents on the western sides of continents in middle and high latitudes, and from the increasing continentality toward the east. It seems entirely reasonable to assume that a similar distribution of vegetation and temperature during the Eocene resulted from the same controlling factors, and to conclude that land and sea relations, as well as planetary circulation, were essentially like those of to-day. Brooks shows a map of part of Europe, after F. von Kerner, with a similar trend of Tertiary isotherms, based on wholly different evidence."

Like Berry, whose view was mentioned above, Chaney [1]) concludes that all paleobotanic data plead against the idea of a shifting of continents and poles since the Eocene, and his final verdict is summed up as follows: "The pattern of land vegetation around the Northern hemisphere indicates that the factors controlling air and water circulation have been essentially the same since the Eocene — that North America and Eurasia have stood in their present positions since the dawn of the Cenozoic. Evidence of land plants of earlier times, less completely known and interpreted, seems also to refute the hypothesis of continental drift. Forrests under compulsion of

1) Chaney, op. cit. p. 486.

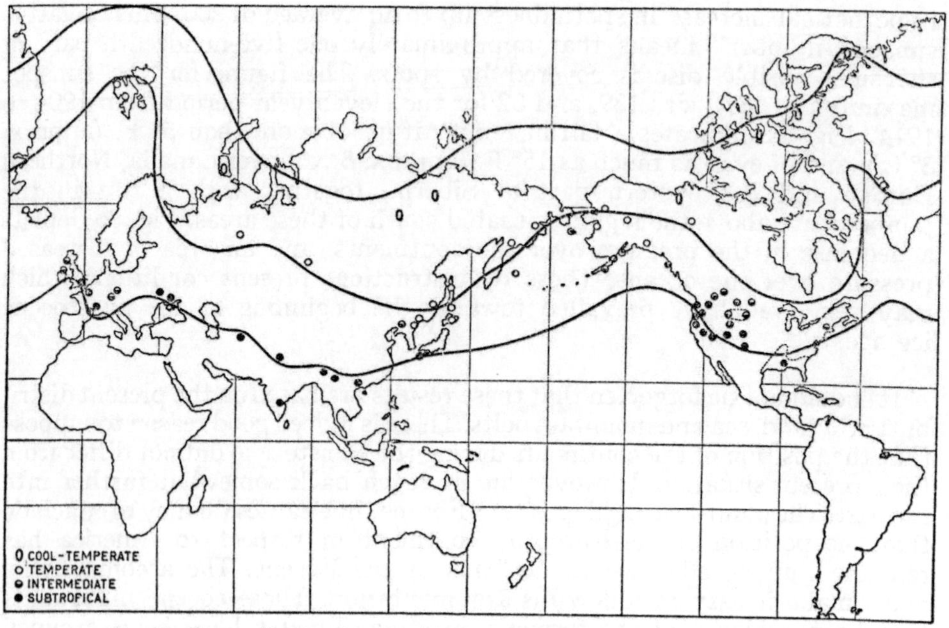

Fig. 87. Eocene "isoflors" in the Northern Hemisphere (From R. W. Chaney).

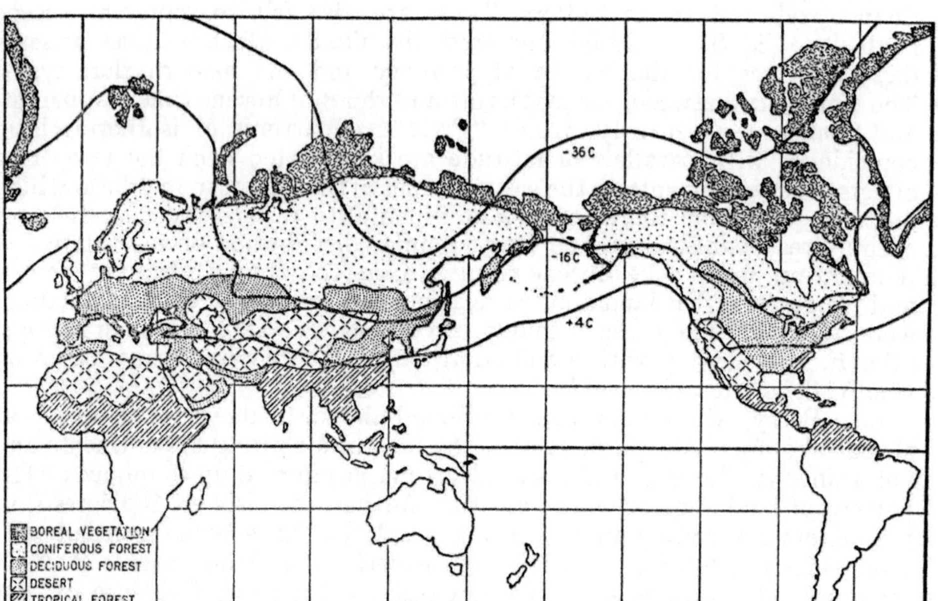

Fig. 88. The distribution of modern vegetation and January isotherms in the Northern Hemisphere (From R. W. Chaney).

climatic change, rather than the continents on which they live, appear to have been the wanderers during the history of life upon the earth."

This brings us to the problem of the position of the upper-paleozoic ice-sheets. Chaney's results, together with Brooks' noteworthy investigations, make it clear that the unusual distribution of upper-paleozoic and even more ancient glaciations could in all probability be explained without continental drift if only accurate paleogeographic maps could be reconstructed for those distant ages.

However, some authors regard the distribution of the tillites in the Upper-Paleozoic as one of the most important arguments in favour of the hypothesis that vast changes affected the position of the continents and poles. It was explained above that the reconstruction of climatological zones, based on the present situation, as put forward by Köppen and Wegener, was a fundamentally wrong one for several reasons. For, to begin with, these investigators assume that the present conditions are the normal ones; but — as Coleman remarked — "that is far from being the case." In the second place, they do not explain — nay, even obscure — one of the most prominent climatic features, viz. the periodicity of abnormal climates. There can be no doubt about the occurrence of this last phenomenon, for, as shown above, it takes place in conjunction with other periodical processes in the earth's crust.

As mentioned previously, the abnormal climates may be explained by the combined influence of widely diverging factors, among which, however, the shifting of poles and continents has played no part.

I would now like to examine a few other points, of which no account was taken so far. Astronomical data do not support the view that the position of the earth's rotation-axis has undergone any significant change, nor does the astronomer find it possible to suggest a cause for such a hypothetical occurrence. Moreover, the comparatively rapid genese of glaciation would require equally rapid hypothetical changes, but since the earth is a gyroscope, and thus has a powerful tendency to keep the axis of its rotation constant, any sudden alteration in its direction would only bring about a world-wide catastrophe [1]). Historical geological data prove that nothing of the kind has ever happened.

On the other hand, it might be assumed that the position of the earth's crust has undergone some changes in respect to the core. This would result in a different distribution of the continents in respect to the axis of rotation, and the latter's position would then undergo a slight alteration owing to this process. It was this type of "shifting of poles" that was used by Köppen and Wegener when constructing their paleogeographic maps .Milankovitch published a mathematical treatise on this subject, showing that the North Pole lay in the vicinity of the Hawaiian Islands in the Paleozoic, and that since then it had migrated to its present position via Alaska.

Mathematical calculations should always be welcomed when dealing with geological problems, for they frequently sift faulty arguments and

---

1) Coleman, op. cit. 1926, p. 263.

incorrect suppositions .Yet an entirely accurate and un-impeachable mathemathical calculation may sometimes lead to wrong conclusions. As Huxley observed [1]: "Mathematics may be compared to a mill of exquisite workmanship, which grinds you stuff to any degree of fineness; but nevertheless what you get out depends upon what you put in; and as the finest mill in the world will not extract wheat-flour from peacods, so pages of formulae will not get a definite result out of loose data". The premisses of Milankovitch's calculations are such that its results can be said to be the reverse of what might be termed definite proof. For one of the assumptions is that no change has affected the sialic shell, and another that an activity, referred to by Milankovitch as a "Polfluchtkraft", was in existence, etc. Yet these conjectures are so questionable that his conclusions may be said to lose all value, as was shown at great length by Schwinner.

I would like to mention a few additional points at this stage. The fusion of the continental blocks into a single, large primordial continent, which is known as Pangaea, would provide an adequate explanation, in the opinion of the drift hypothesis, of the present widely-scattered remnants of the upper-paleozoic glaciations, for they would have united as one, extensive ice-cap in the South-Pole. Yet several authors, including Coleman, doubt whether such an extensive land mass would have acted favourable on the formation of an ice-cap. The reconstruction of a continent such as Pangaea was discussed exhaustively by Brooks, who examined to what extent climatic evidence might be said to plead for the idea of continental drift. One of this investigator's conclusions was that the relative ages of the glacial deposits in the southern hemisphere are incompatible with Wegener's view of moving poles. Another point is that tillites occurred in the northern hemisphere, particularly in North America, North-Western Siberia, (where the ice probably radiated from the existing Kara Sea Massif) and Central Asia (where it descended from the region of Tarim). It should be remembered that the existence of an enormous ice-cap in Asia during the Pleistocene only became known quite recently. The Paleozoic of Asia may consequently hold more surprises in store for us in this respect! The presence of tillites has been reported in numerous other areas. Some may have been brought thither by local mountain glaciers or floating ice-masses. Others, such as those existing in Texas, are probably tectonic breccias, but the so-called Squantum-tillite near Boston is so thick and wide-spread that geologists have not hesitated to assume that these fossil moraines very probably constitute remnants of a vast ice-sheet. In Wegener's reconstruction of Pangaea, the latter is situated in his equatorial rain-zone of Carboniferous time, and in the center of his Permian desert belt. These tillites consequently conflict with the theory of drift, and Wegener therefore attempts to "explain them away tacitly" [2]. Brooks expressed the opinion that the Squantum-tillites might be explained without a shifting of poles and continents, and assumes that they were formed under the influence of a cold sea, extending from the Arctic region as far as the eastern part of America [3]. Brooks

1) *Fide* Knowlton. p. 565.

2) Brooks, p. 266.

3) It is far from certain, however (see fig. 89), whether a cold sea existed in North-America, which would thus explain the presence of the above Squantum-tillites. Some geologists have

devoted a whole chapter to the causes of the abnormal climate of Upper-Paleozoic time (these are dealt with at great length in his Chapter XV), and also showed that a possible westward drift of both the American continents in respect to Europe could not affect his solution in any way, though the change as regards the position of other continents would only augment the problem's difficulties. As a matter of fact, Brooks not only pointed to the defects of the hypothesis of drift, but also attempted to furnish a constructive and extensively argumented explanation of the question of geological climates, in which "epeirophorese" played no part whatever.

We are living in an age of ice-caps. The distribution of land, sea and mountain chains, etc. are responsible for the position of an ice-cap as seen above. It was likewise shown that the site of the pleistocene ice-caps can be understood without a hypothetical shifting of continents. The emplacement of the upper-paleozoic ice-caps, too, could probably be explained if we had all available data as regards the distribution of physiographic factors in the distant past at our disposal.

My own conclusion may be said to concur with that of Schuchert, which the latter expressed 20 years ago. When asked what had brought about the upper-paleozoic glaciations, this author replied [1]: "Certainly never the impossible Wegenerian hitching together in close embrace of the present masses of South America, Africa, India, and Australia. In this hypothesis I see only a drunken sialic upper crust hopelessly floundering upon the sober sima. Let us return to a tangible geology. We all now know that the earth underwent one of its greatest crustal disturbances during the Carboniferous, beginning late in the Lower Carboniferous (toward the close of Viséan and Chesterian time), attaining a first climax in middle Upper Carboniferous time, and a second maximum early in the Permian. The supercrust was undergoing one of its periodic revolutions, and the world was then as scenically grand as it is to-day. It is in the youthful topography, the enlarged continents, and the peculiar connections of the continents that seemingly are to be sought the reasons for the Permian ice-age." To this he added, some nine pages further on: "If it can be shown that Australia was connected in Permian times with Antarctica, then this holding in of so much of the cold waters of the Antarctic Ocean, combined with moist climates in the southern hemisphere and the general highland condition of so much of the world in early Permian time, will be the explanation for the peculiar position of the continental ice-masses of the southern hemisphere".

Schuchert and Willis later illustrated the supposed distribution of land and sea, as well as that of cold and warm ocean currents, and tried to explain the upper-paleozoic glaciation without the occurrence of alterations in the present position of continental blocks. These authors, like Brooks, are of the opinion that a solution must take into account the existence of former land

drawn a sea-arm in this region, but others indicate land in this area instead of water. This may be enough to show that Brooks' explanation, too, should be taken *cum grano salis*. Still, no other alternative is left us, for it would be impossible to construct a paleogeographic map which would not be subject to criticism, or which would not at times even give rise to severe controversy.

[1]) Schuchert, 1923, p. 1081.

connections between Africa and Australia, and between Africa and South-America. Many investigators have based similar conclusions on other grounds (i.e. on various geological arguments, and the distribution of land plants, animals and marine faunas).

The preceding chapter explained that though each reconstruction of transoceanic land-bridges retains a purely hypothetical character, the idea that land-bridges existed in former times should not be wholly discarded. A "not impossible" reconstruction will be found in fig. 89.

At some time in the future, when science will have progressed still further, it will perhaps be possible for geologists to draw a complete and accurate

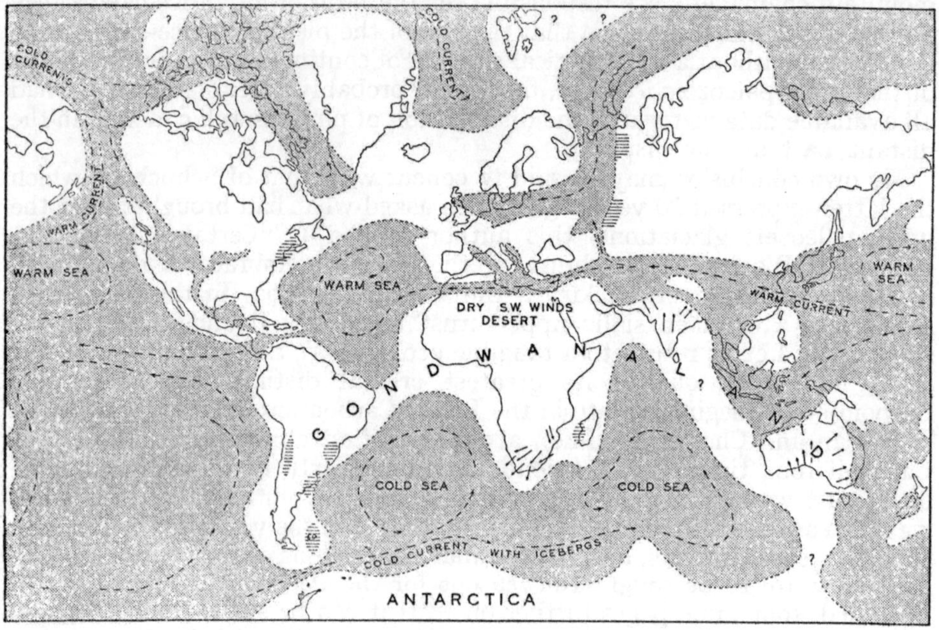

Fig. 89. Hypothetical reconstruction of geographic conditions during the upper-paleozoic ice ages. Glaciated regions are indicated by horizontal shading and small arrows; the latter show the direction of ice-flow in South-Africa and Australia.

paleogeographic and paleoclimatologic map of the Upper-Paleozoic. A more profound knowledge of the geological history of the oceans will perhaps have been acquired by that time. Until then, however, a "not impossible" reconstruction such as that appearing in fig. 89 will have to replace a more detailed and satisfactory explanation of the remarkable glacial remnants of upper-paleozoic time.

The restricted influence of cosmic factors

The influence of cosmic phenomena was met with twice in the foregoing paragraphs. Huntington and Visher, in their „solar cyclonic hypothesis", went so far as to regard a great increase in sunspots as the only primary factor capable of bringing about ice-ages. Yet there are two objections to this theory: (1) that a periodicity of the fluctuation of solar radiation, such as is sought when attempting to explain the earth's abnormal climates, is, of course quite unknown, and that its existence has never been made acceptable; (2) that abnormal climates, as previously stated, are connected with internal terrestrial processes (mountain-building, magmatic cycles, etc.). The influence of sunspots on subcrustal processes and its effect on epochs of folding, mountain-building, regressions, etc. would therefore have to be shown first. Huntington and Visher furnish tables purporting to show the apparent relation of an increase in spottedness to an accompanying increase in tectonic earth-quakes, and retrace said relation via changes in the atmospheric pressure. They derive the hypothetical and major periodicity of sunspots from another theoretical process, in which the sun is supposedly affected by changes in the position of other heavenly bodies. Their combined views thus attribute a direct influence on the climate, and an indirect one on subcrustal events to cosmic phenomena. An illustration of their argumentation appears in fig. 90. However, this figure also includes the effect of geographical factors, since a hypothesis which attempts to explain everything solely by means of cosmic phenomena seems to me to be hopelessly incompatible with present geological data.

Nevertheless, it may be that the influence of cosmic phenomena, as represented in this scheme, is not sufficiently restricted. Joly had previously already observed that it was not improbable that the periodicity of subcrustal processes was stimulated by a cosmic effect and Schwinner subsequently argued that after the supposed disruption of the moon from the earth a resonance effect would have occured as much as eleven times (i.e. the period of the tides would have coincided eleven times with the natural free period of vibration of the earth), and he assumes that this would have caused the periodic phenomena of mountain-building. (fig. 91). Yet, it makes a considerable difference, from a fundamental point of view, whether we restrict the cosmic influence in this manner, or attribute the dominating role to astronomic factors, as in the "solar cyclonic hypothesis."

Besides, even when restricted thus the above astronomical influence can be stated to constitute no more than a hypothetical factor. Prof. Oort recently assured me that the only exceptionally long periodicity established so far in astronomy is the rotation of our galactic system, which revolves in approximately 200 million years. This figure agrees fairly well with that of the major geological rhythm of 250 million years. This congruence is indeed remarkable, but for the moment it is hard to see how the astronomical cycle could be connected with atmospheric and subcrustal processes.

Summary

The Canadian geologist Coleman, who gleaned a considerable knowledge as regards glacial phenomena from direct observation over the entire world, remarked towards the end of his book on Ice-Ages that he did not consider himself competent to propose a hypothesis which would prove to be more satisfactory than the existing ones (it should be noted, however, that he did not hesitate to draw attention to the defects in all the previous theories). Coleman ended, however, with these words: "In my opinion no single cause, as expressed in one of the theories proposed, can accomplish this; and any final solution of the complicated problems involved must come from a conjunction of general and local causes. Some combination of astronomic, geologic and atmospheric conditions seems to be necessary to produce such catastrophic events in the world's history."

I regard this modest view as the only correct one. The problem is undoubtedly a very intricate one. I have attempted to emphasize the necessity of combining the various cooperating factors, and my motives may be summarised as follows (see fig. 90 and 91): an almost world-wide and temperate climate prevailed during such a considerable lapse of time that we may consider it to have constituted the normal climate. Deviations from this normal state of affairs, which produced large glaciations, occur with a periodicity of approximately 250 million years. These are related to the principal precambrian, upper-paleozoic and pleistocene periods of mountain-building that influenced large parts of the globe. Less intensive climatic variations with a shorter periodicity occur during the intervening periods and are probably related to mountain-building of minor geographical significance. Mountain-building is closely related to diverse other groups of phenomena (regressions, changes in the pressure of the earth's crust, magmatic processes, etc.), which point to a common and wide-spread cause: i.e. to the activity of periodical subcrustal processes. The latter might also be thermal cycles, and the fall of temperature might then possibly concur with periods of mountain-building and glaciation. When the alterations in the earth's physiographic aspect (with the possible inclusion of subcrustal cooling) produce local ice-caps, these suddenly begin to expand and assume unusual proportions as a result of the cooling effect exerted by an ice-sheet on the temperature in the surroundings. The special position of ice-caps shows that they are controlled by physico-geographical factors: the distribution of land and sea, mountains, cyclone routes, etc. Nevertheless, the influence of astronomical factors cannot be denied. To begin with, it may be possible that an increase in the amount of sunspots finds expression in the atmosphere, i.e. in a decrease in the temperature and pressure above certain continental areas. And in the second place, the division of glaciations into ice-ages and interglacial stages may probably be ascribed to cosmic influences, and may probably be derived from changes in the obliquity of the ecliptic, the variable eccentricity of the earth's orbit and the precession of equinoxes.

All these effects might be combined in two ways. One would be to assume (fig. 91) that the periodicity of the abnormal climates is activated primarily by internal terrestrial processes. Their effect would suddenly increase to a considerable degree owing to the cooling influence of an ice-sheet. The influence of cosmic phenomena, moreover, would increase the effect still further, and these phenomena would explain the appearance of interglacial stages. The subcrustal processes may possibly also be stimulated by cosmic phenomena.

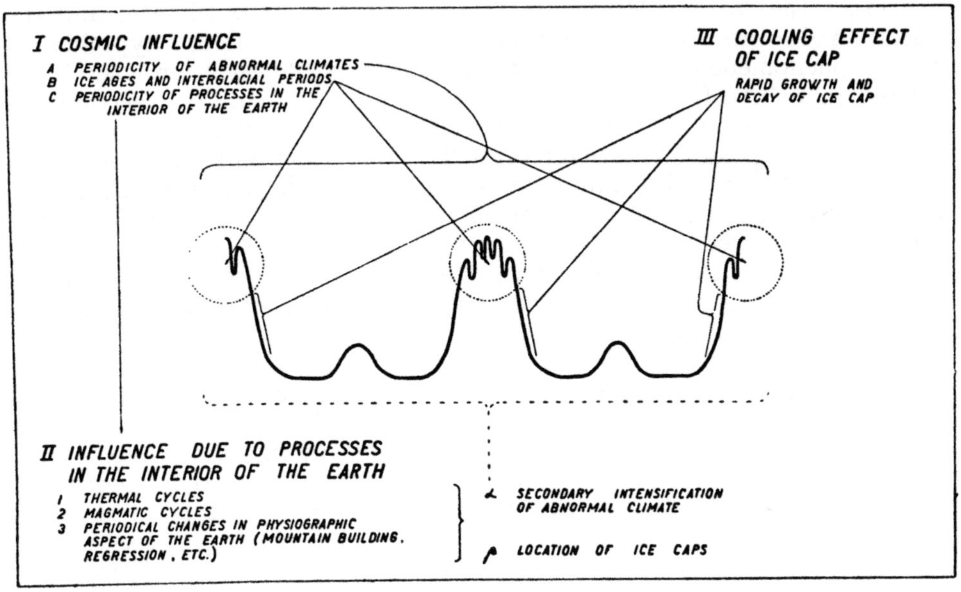

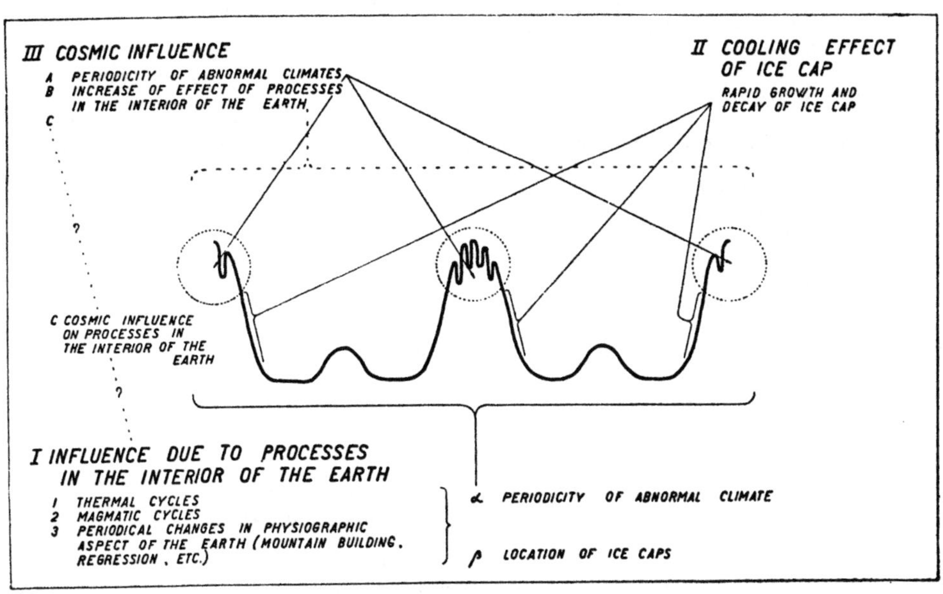

Fig. 90 and 91. Schematic representation of two possible explanations of glaciation and ice-ages.

The alternative (fig. 90) would be that the abnormal climates and the initiation of events in the globe's interior are the direct result of a hypothetical cosmic phenomenon. The subcrustal processes would therefore merely be related in point of time to abnormal climates, and would exert a mere secundary effect on the latter, at the most. One cosmic phenomenon is in fact known to possess a periodicity of the same order as that found in the major periodicity of 250 million years on earth, viz. the rotation of the Galactic System. It is not yet clear, however, how this astronomical period might be connected with subcrustal events.

In my opinion the solution of this tangled problem must be sought in one of these contingencies. The question, at present, is whether the primary cause must be attributed to internal terrestrial processes or to cosmic factors. Some may incline to the first and others to the second alternative when looking for a probable explanation of the periodical occurrence of abnormal climates, but it cannot be denied that it is impossible — or expressed more optimistically, not yet possible — to determine the primary cause of this phenomenon. Daly's negative view, which was mentioned at the beginning of this chapter, can thus be said to be correct up to this point. I hope, however, that I have succeeded in showing some of the features of the periodical occurrence of the remarkable phenomena of glaciation and ice-ages.

References

ANTEVS, E. *Maps of pleistocene glaciations* (Bull. Geol. Soc. America, 40, 1929)

BACKLUND, H. G. *On late palaeozoic glaciations in the northern hemisphere* (Uppsala Universitets Arsskrift 1929).

BERRY, E. E. *Climatic significance of arctic fossil floras* (Bull. Geol. Soc. America, 40, 1929).

BROOKS, C. E. P. *Climate through the ages* (London, E. Blenor 1926).

BUCHER, W. H. *Deformation of the Earth's Crust* (Bull. Geol. Soc. America 1939).

CHANEY, R. W. *Tertiary forests and continental history* (Bull. Geol. Soc. America, 51, 1940).

COLEMAN, A. P. *Ice ages* (Macmillan Co 1926).

COLEMAN, A. P. *Ice ages in the geological column* (Bull. Geol. Soc. of America, vol. 50, 1939).

DACQUÉ, E. *Grundlagen und Methoden der Paläogeographie* (p. 376–476. Die Paläoklimatologie 1915).

DACQUÉ, E. und WEGENER, A. *Paläogeographie* (in Enzyklopädie der Erdkunde, 1926).

DALY, R. A. *The changing world of the ice age* (1934).

DARRAH, W. C. *Principles of Paleobotany* (1939).

DU TOIT, A. L. *Our wandering continents* (Oliver and Boyd 1937).

EVERDINGEN, E. VAN, *Verklaring van geologische klimaten* (Tijdschr. Kon. Nederl. Aardrijkskundig Genootschap 57, 1940).

FORREST, H. E. *The Atlantean Continent. Its bearing on the great ice age and the distribution of species* (Witherby, London 1933).

GERTH, H. *Die Tertiärfloren des südlichen Südamerika und die angebliche Verlagerung des Südpols während dieser Periode* (Geolog. Rundschau. 32, 1941).

GRIGGS, D. A. *Theory of mountain building* (Americ. Journ. of Science 1939).

HOBBS, W. H. *The glacial anticyclones* (1926).

HOLMES, A. *A contribution to the theory of magmatic cycles* (Geol. Magazine 63, 1926).

HOLMES, A. *The age of the earth* (1937).

HUMMEL, K. *Welteislehre und Geologie* (Natur und Volk, 69, p. 129, 1939).

HUNTINGTON, E. and VISHER, S. S. *Climatic changes* (Yale Univ. Press 1932).

JOLY, J. *The surface history of the Earth* (1925).

JOLY, J. *The theory of thermal cycle* (Gerlands Beiträge zur Geophysik 1928).

KERNER-MARILAUN, F. *Paläoklimatologie* (Berlin, Borntraeger, 1930).

KNOWLTON, F. H. *Evolution of geologic climates* (Bull. of the Geol. Soc. of America, vol. 30, 1919).

KÖPPEN, W. und WEGENER, A. *Die Klimate der geologischen Vorzeit* (Berlin, Borntraeger 1924).
MILANKOVITCH, M. *Théorie mathématique des phénomènes thermiques produits par la radiation solaire* (Paris, Gauthier-Villars 1920).
MILANKOVITCH, M. *Säkulare Polverlagerungen* (Handb. d. Geophysik I, Lief. 2, 1933).
MILANKOVITCH, M. *Astronomische Mittel zur Erforschung der Erdgeschichtlichen Klimate* (Handb. d Geophysik IX, 1938).
NORIN, F. *An occurrence of late palaeozoic tillite in the Kuruk-tagh mountains, C. Asia* (Adress G. F. Stockholm Meeting 1929).
OESTREICH, K. *IJstijd en actualisme. Opmerkingen naar aanleiding van de Verhandeling van Dr. C. G. S., Sandberg: Ist die Annahme von Eiszeiten berechtigt?* (Tijdschr. Kon. Nederl. Aardr. Genootschap 58, 1941).
OSBORN, H. F. *Ancient Vertebrate life of Central Asia* (Livre Jubilaire. Soc. Geol. de France II, 1930).
RAMSAY, W. *The probable solution of the climatic problem in geology* (Geol. Magazine 61, 1924).
RUEDEMANN, R. *Climates of the past* (Geologie der Erde, North America 1939).
SALOMON-CALVI, W. *Die Permokarbonischen Eiszeiten* (Leipzig, Akad. Verlag 1933).
SANDBERG, C. G. S. *Ist die Annahme von Eiszeiten berechtigt?* (Ed. Sijthoff, Leiden 1937).
SAURAMO, M. *The Quarternary geology of Finland* (Bull. Comm. Geol. Finlande 89, 1929).
SCHUCHERT, CH. *The Earth's changing surface and climate*, (in R. S. Lull. The Evolution of the Earth, Yale Univ. Press 1919).
SCHUCHERT, CH. *The paleogeopraphy of Permian time in relation to the geography of earlier and later periods* (Proceed. Second Pan Pacific Sci. Congres, vol. 2, 1926).
SCHUCHERT, CH. *Gondwana landbridges* (Bull. Geol. Soc. of America, vol. 43, 1932).
SCHWINNER, R. *Sind grosse Polverschiebungen möglich?* (Gerlands Beiträge zur Geophysik 43, 1934)
SCHWINNER, R. *Lehrbuch der physikalischen Geologie* (I, 1936).
SIMPSON, G. C. *World climate during the Quarternary period* (Quart. Journal Royal Meteorol. Soc. 60, 1934).
SOERGEL, W. *Die Vereisungskurve* (1937).
SPITALER, R. *Die sommerliche and winterliche Bestralung während der quartären Eiszeit* (Meteorolog. Zeitschrift 1939).
TEICHERT, C. *A tillite occurrence on the Canadian Shield* (Report of the fifth Thule Expedition, Vol. I, Nr. 6, 1937).
UMBGROVE, J. H. F. *Oorzaken van IJstijden* (Natuurkundige Voordrachten, Mij. Diligentia, 's-Gravenhage 1941).
VENING MEINESZ, F. A. *The determination of the Earth's plasticity from the postglacial uplift of Scandinavia* (Proc. Kon. Acad. v. Wetensch. Amsterdam, vol. 40, 1937).
WAGNER, A. *Klimaänderungen und Klimaschwankungen* (1940).
WANLESS, H. R. and SHEPARD, F. P. *Sea level and climatic changes related to late Paleozoic Cycles* (Bull. Geol. Soc. of America, vol. 47, 1936).
WATERSCHOOT VAN DER GRACHT, W. A. J. M. VAN, *The Permocarboniferous exotic boulders in the so-called "caney shale" in the northwestern front of the Ouachita mountains of Oklahoma* (Journal of Geology, vol. 39, 1931).
WILLIS, B. *Isthmian links* (Bull. Geol. Soc. of America, vol. 43, 1932).

Chapter VIII

THE RHYTHM OF LIFE

"....changing environmental conditions stimulate the sluggish evolutionary stream to quickened movement".

(R. S. Lull)

Introduction

Environmental factors have had a decided influence on the evolution of Life. It can hardly be an accident, for instance, that geological phenomena such as regression and the laramide mountain-building towards the close of the Mesozoic, should have corresponded with the equally important adaptive radiation of placental mammals. Then, too, can it be mere chance that the Psychozoic — which received its name from the fact that Man, the specialist of spiritual and intellectual differentiation, began to extend his supremacy over the world at that time — coincides with an exceptionally intensive phase of mountain-building and glaciation?

Many more examples could be given, but we will confine ourselves to the above two. In one of his well-known treatises, Matthew emphasized the influence of the climate on organic evolution, regarding it as the most important environmental factor to effect biological changes. This significant primary role was attributed by Szalai to the various manifestations of mountain-building, and by Wilser to the varying intensity of solar radiation. Other factors, too, have undoubtedly played a more or less important part, but it will be clear that it is scarcely possible to estimate each influence separately, since all are related to one another to a greater or lesser degree. It would lead us too far to examine the absorbing biological problems inherent to the changes to which the geneplasm and hereditary constitution were obviously subjected. For the moment we merely wish to point to the frequent concurrence of changes in the physical and organic world. As Lull remarked in the closing lines of his chapter on "The Pulse of Life": "Thus time has wrought great changes in earth and sea, and these changes acting directly or through the climate, have always found somewhere in the unending chain of living beings certain groups whose plasticity permitted their adaptation to the newly arising conditions.

"The great heart of nature beats, its throbbing stimulates the pulse of life."

The evolution of the fauna was not only affected by many changing terrestrial factors, but also adapted itself to the evolution of the flora. The boundary between the Paleozoic and the Mesozoic lies on a higher stratigraphic level than that between the Paleophytic and Mesophytic, and the Cenophytic, too, began at an earlier date than the Cenozoic! The effect of external changes on the evolutionary stream of life might therefore *a priori* be expected to reflect itself more distinctly in the flora than in the fauna. That this influence on the flora is in fact very evident will be shown in the following paragraphs.

The periodic differentiation of the flora

The flora appears to have been entirely uniform during the Devonian, judging by what is known of its distribution during that period. An almost equable land-flora also characterizes the Lower-Carboniferous (fig. 92). This uniformity became far less pronounced in the Upper-Carboniferous, and separated into four more or less well-defined provinces (fig. 93). One of these flora-types is called the Atlantic-Chinese by Seward, the Arcto-Carboniferous by Darrah and the Eurameric by Jongmans. It includes such well-known genera as *Calamites, Cordaites, Lepidodendron, Sigillaria,* as well as many *Pteridospermae* and ferns, and covers parts of Eastern North-America, Greenland, Europe and North-Africa. This flora-type extended eastwards as far as Korea and Sumatra during the middle division of the Upper-Carboniferous, mingling with plants of the Gigantopteris flora [1]). The latter is also known as the Cathaysia flora and developed in China, but it has also been found — remarkably enough — in Texas and Oklahoma! The third botanical province comprises the Glossopteris flora, with as its most outstanding genera: *Gangamopteris, Glossopteris* (*Vertebraria*), *Phyllotheca, Gondwanidium* and *Dadoxylon,* to which should also be added such types as *Lepidodendron* and *Psaronius.* The latter are frequently found among the Eurameric flora. No sharp line can be drawn between the different provinces, and a particularly strong mixture of Eurameric and Glossopteris plants seems to exist in Rhodesia [2]). Jongmans recently reported that he had come across such elements as the Cathaysia and Glossopteris flora among upper-paleozoic fossils in New-Guinea [3]).

The same author restricted the name Gondwana flora to the Glossopteris plants of the southern continents, and speaks of the more or less analogous plant-finds around Angaraland (Russia, North Central-Asia and East-Asia), as the Angara flora. Seward, who refers to the last flora-type as the Kusnezk flora, draws attention to the fact that fossil plants very similar to the Kusnezk flora were discovered in Permian strata (in the uppermost part of the Lower-Permian) in Arizona!

Thus four distinct botanical provinces can be observed among the flora of the Upper-Carboniferous and the Permian. This differentiation, which corresponded with tremendous mountain-building and important changes in the distribution of land and sea, together with exceptionally wide-spread

1) EC in fig. 93.
2) Seward, p. 246. EG in fig, 93.
3) CG in fig. 93.

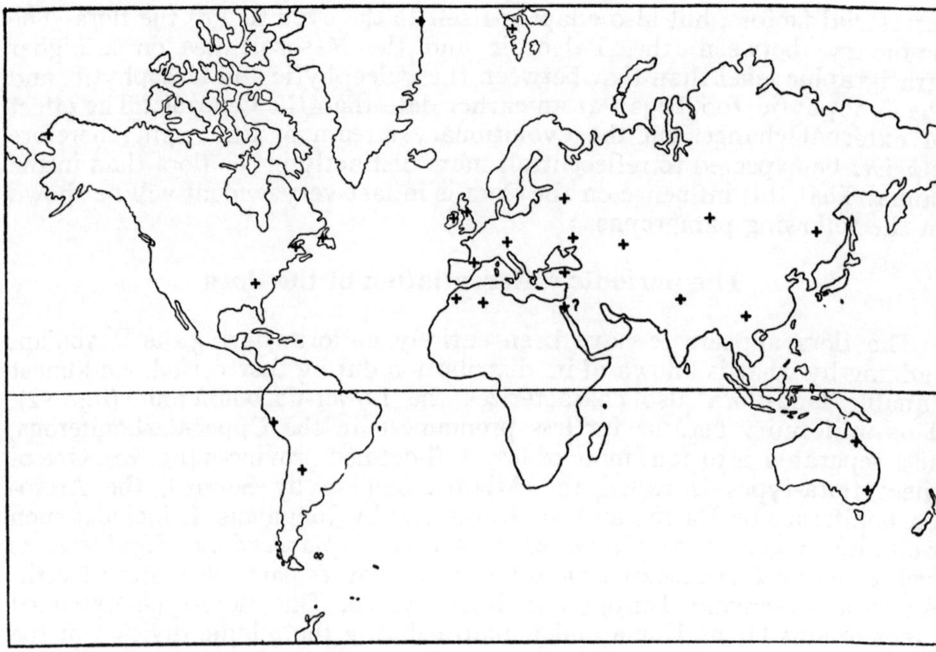

Fig. 92. The distribution of a selected number of localities where the cosmopolitan flora of the Lower-Carboniferous has been found (After A. C. Seward).

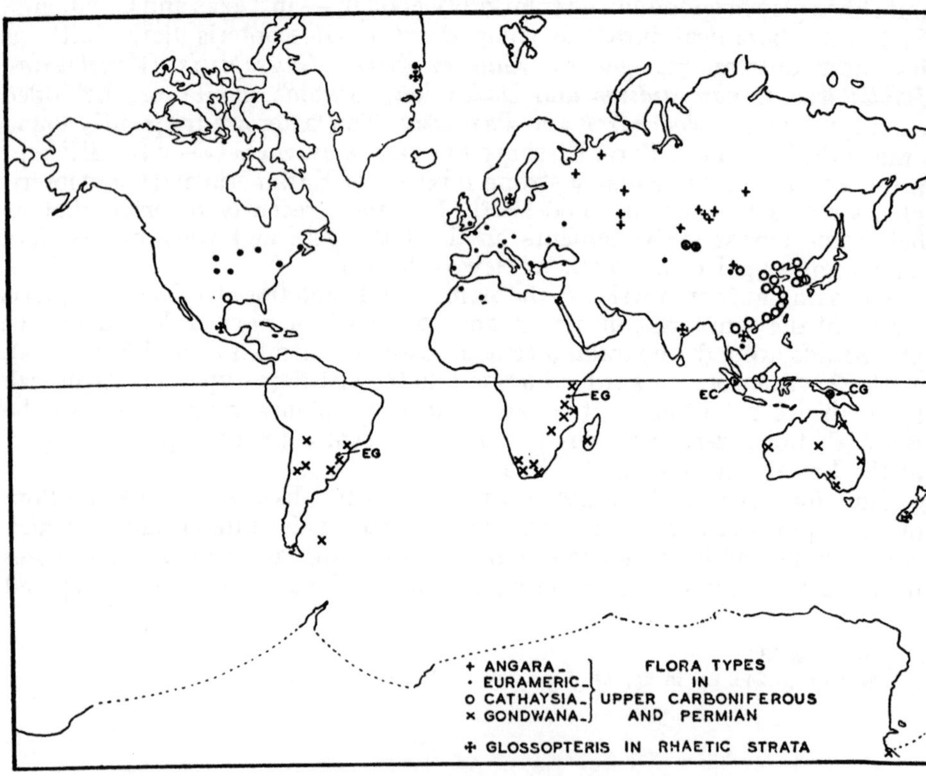

Fig. 93. Upper-paleozoic flora types.

glaciation (which, according to investigations by Du Toit, David and Süssmilch, began in the Lower-Carboniferous and continued with a few phases up to the Lower-Permian) can hardly be said to be due to mere chance. On the contrary, the important alterations in the earth's physiographic aspect during the Upper-Paleozoic probably stimulated the differentiation of the flora.

It is therefore especially noteworthy that the "levelling" process to which the climate was subjected during the Mesozoic (see Chapter VII) should have been accompanied by a cessation of the extreme differentiation of botanical provinces. For it is well-known that after the upper-paleozoic glaciation the land flora developed into what might almost be called a world-wide uniformity during the Triassic, Jurassic, Cretaceous and even up to the Tertiary, though this uniformity was distinctly less accentuated during the last period. This uniformity was described by Seward [1]), and Darrah [2]), too, observes: "Perhaps the most remarkable feature of the early Mesozoic flora is its homogeneity, both in space and in time. The flora does not vary greatly from the middle Triassic to almost the end of the lower Cretaceous. Of course, the specific content of the floral succession changes, but the generic content is strikingly constant".

A sudden and as it were explosive evolution of the Angiospermae may indeed be observed in the Upper-Cretaceous. In this case, too, it could hardly be claimed that the fact that this intensive evolution coincided with a phenomenon of world-wide importance, — i.e. one of the largest transgressions that have ever occurred in geological history — is merely a question of chance. Seward [3]) consequently assumes that these vast changes in the physiographic aspect of our globe had an undeniable influence on the evolution of plant-life on the continents.

The wide-spread uniformity of the climates had ceased to exist in the Lower-Tertiary (i.e. after the period of laramide mountain-building) and was immediately followed by a zonal differentiation of the flora, which — like that observed in the Miocene [4]) — was roughly arranged according to latitudes. This differentiation was far less pronounced than it is at present, for the latter has resulted from the strong climatic changes since the Pleistocene, and these were responsible for the distribution of our plant-life into the actual botanical provinces. The steadily expanding ice-sheets subsequently drove the widely-scattered, circumpolar miocene-pliocene flora of Europe, Asia and North-America in a southerly direction, separating them into smaller units as the ice pushed forward. The plants of the northern hemisphere survived where they were able to migrate southwards, and thus found it possible to return north and repopulate the glaciated areas as soon as the ice began to retreat. Darrah [5]) writes in this connection: "The Miocene and early Pliocene floras were practically circumpolar in distribution and generally spread over the northern hemisphere with the zonations already indicated. The oncoming of colder conditions probably forced the

---

1) Seward 1931, p, 332–334, 368, 371, 404, 406, 454–455.

2) Darrah, p. 154.

3) Seward, p. 404.

4) Darrah, p. 187, 192, 193.

5) Darrah, p. 195, 196.

plants southward. Presumably there were three principal avenues for migration or escape, and these were determined by the prevailing direction of the mountain systems. One avenue of migration was along the lowlands of eastern Asia along the great valley systems and the coastal plains. It has been suggested that the richness of the existing flora of China is due to the intermingling or persistence of northerly species with those already established.

"In North-America the north-south trend of the mountain ranges permitted fairly free migration, and many species of plants which came by this route and those which are relict in place, enrich the American flora. This is the explanation for the marked similarity between the floras of eastern North America and eastern Asia, first observed by Asa Gray, and since demonstrated by many others.

"This resemblance extends not only to many genera with closely related species, but to a considerable number of many identical species. A study of the flora of the Island of Yezo, which lies to the north of the main island of Japan, discloses that more than twenty-six per cent of its plants are found also in North America.

"The third migration route for the floras of the north, as they were forced southward by the increasing cold, was down the Scandinavian Peninsula and adjacent areas into central and western Europe, but here the physiography is very different from that of North-America and eastern Asia. The great mountain systems, i.e. the Alps, Pyrenees, Carpathians, Balkans, and Caucasias all trend east and west, and thus prevent further southward migration. Many of the northerly Miocene and Pliocene plants actually reached western Europe, as is shown by their presence in the Reuverian flora, but, as the cold increased, they were forced against the mountain barriers and perished.

"The Teglian flora of the late Pliocene is very different from the Reuverian flora, and apparently had two centers of origin, one from Scandinavia, and the other from the mountains of central Asia. It was a cold-temperate assemblage, and hence when it was pushed southward by the advancing cold, much of it was able to survive."

Should this prove to give a satisfactory explanation of the distribution of the pleistocene and actual flora types we might enquire wether the distribution of flora types in the Upper-Paleozoic were not, *mutatis mutandis*, due to fundamentally analogous circumstances. It would take us beyond the scope of the present chapter, however, to examine this interesting biogeographical question.

The most important point of all, as far as we are concerned, is that the two major periods of strong differentation of plantlife correspond with two major periods of mountain-building and glaciation of the Upper-Paleozoic and Pleistocene.

References

BARRELL, J. *Influence of Silurian-Devonian climates on the rise of air-breathing vertebrates.* (Bull. Geol. Soc. of America, vol. 27, 1916).

BARRELL, J. *Probable relations of climatic change to the origin of the tertiary ape-man* (Scientific Monthly 1917).
BARRELL, J. *Relation between lapse of time and organic evolution* (Bull. Geol. Soc. America, vol. 28, 1917).
COTT, H. B. *Adaptive coloration in animals* (Methuen & Co, 1940).
DARRAH, W. C. *Principles of Paleobotany* (Ed. Chronica Botan. Cy. Leiden 1939).
JONGMANS W. J. *Die Kohlenbecken des Karbons und Perm im U.S.S.R. und Ost-Asien* (Jaarversl. Geol. Stichting Heerlen over 1934–1937, 1939).
JONGMANS, W. J. *Beiträge zur Kenntniss der Karbonflora von Niederländisch Neu-Guinea* (Ibid. 1940).
LULL, R. S. *The pulse of Life* in "*The evolution of the Earth*" (Yale Univ. Press 1919).
MATTHEW, W. D. *Climate and evolution* (N.Y. Acad. Science vol. 24, 1915).
NICHOLS, G. E.; WOODRUFF, L. L.; PETRUNKEVITCH, A.; COE, W. R.; WIELAND, G. R; DUNBAR, C. O.; LULL, R. S. and HUNTINGTON E., *Organic adaptation to environment* (Yale Univ. Press 1924).
SEWARD, A. C. *Plant life through the ages* (Cambridge Univ. Press 1931).
SZALAI, T. *Der Einfluss der Gebirgsbildung auf die Evolution des Lebens* (Palaeontologische Zeitschrift Bd. 18, 1936).
WILSER, J. L. *Lichtreaktionen in der fossilen Tierwelt* (1931).
ZIMMERMANN, W. *Die Phylogenie der Pflanzen* (Ed. Fischer, Jena 1930).

Chapter IX

THE PULSE OF THE EARTH

"And the secret of it all is in the heart of the earth, forever invisible to human eyes".

(R. A. Daly)

Introduction

The preceding chapters showed time and again, that the different groups of phenomena are genetically related to one another, and that they have a common, deeper cause in a figurative as well as a literal sense, to which we referred repeatedly by means of such neutral designations as "processes in the substratum", "changing conditions in the interior of the earth", "deep-seated forces", "pulsating rhythm of subcrustal processes", etc.

These vague descriptions designate the paramount source of all subcrustal energy, which manifests itself with a periodicity observed in a whole series of phenomena in the earth's crust and on its surface, viz. the alternating decrease and increase of compression of the earth's crust and the closely related epochs of folding, the process of mountain-building and submersion of borderlands, the periodic formation of basins and dome-shaped elevations, the magmatic cycles and rhythmic cadence of world-wide transgressions and regressions, the pulsation of the climate, and — lastly — the pulse of Life.

Chronological relation of periodic events

The object of this last chapter is to recapitulate the above phenomena, and to give a condensed synopsis of their correlation. A diagrammatic view of the preceding events will be found in Table II. The latter finds its explanation in fig. 94, which contains a schematic survey of the chronological relation of the periodical processes of one cycle. An arbitrary lapse of time has here been subdivided into five stages, the division being based on a certain chronological order represented by A, B, C, D and E, and this is followed by a new cycle (A^1 etc.).

A. This first stage is marked by a world-wide regression of the epicontinental seas, resulting either from the elevation of the continents, or the subsidence of the ocean-floors, or even — and this seems the most probable course of events — from the simultaneous but opposed movements of both the continents and the ocean-floors. A period of intensive erosion ensues on the continents, and the climatological zones begin to show a greater differentiation than they did previously. This period is also characterized by an increased compression of the crust, which consequently buckles in geosynclinal areas. One of the consequences of this event is that the contents of the geosynclines become folded, and this is in turn accompanied — and later succeeded — by the intrusion of acid batholiths. A simultaneous phase of movement generally occurs in those basins which were already in existence at that time. We are at present concerned with the surface of the earth in its consolidated state. The hypothetical processes responsible for the origin and arrangement of continental bucklers and oceanic depressions during the early Precambrian will not be considered here (cf. Chapter VI).

B. The first stage is followed by a period of decreasing crustal compression. By this time the processes of folding and regression have already attained their full effect. The folded belts now emerge as mountain chains. Basins and new geosynclines begin to form. Dome-shaped elevations originate, and the sea-level begins to move in an opposite direction (positive movement). The interrupted line in fig. 94 represents a period of only slight regression, and the dot-dash line and full drawn graph in the same column depict periods of extensive and exceptionally extensive regression respectively.

C. The mountain belts have grown to their full height, and the relief of

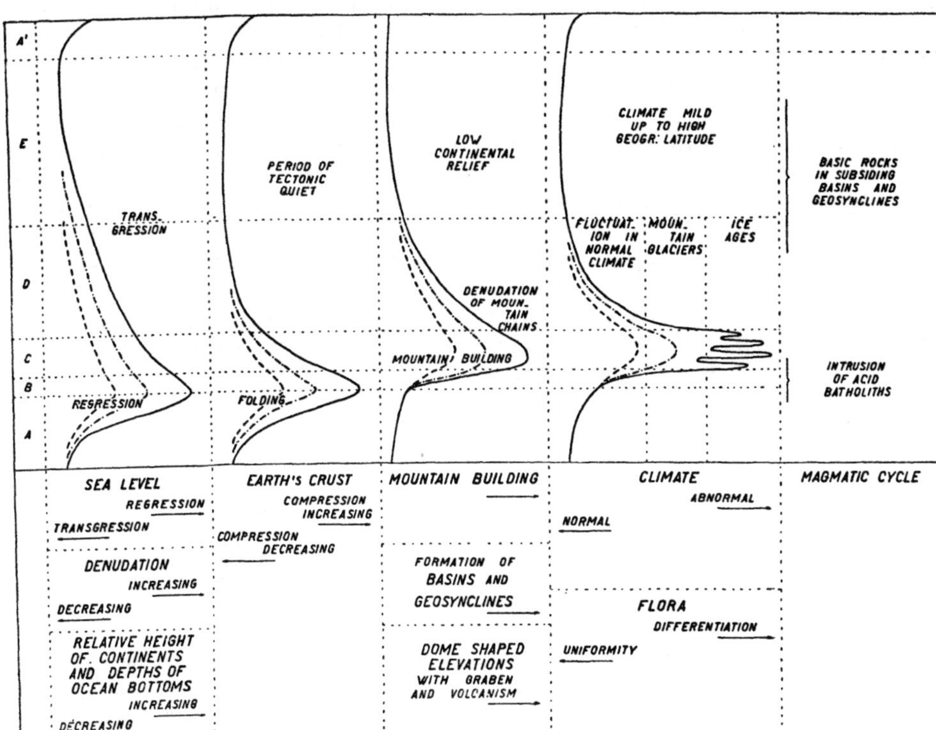

Fig. 94. Graphs showing the chronological relation of the various periodic phenomena.

the continents is very accentuated during this stage. The formation of dome-shaped elevations, rift-valleys and contemporaneous volcanism are now completed. The newly formed basins and geosynclines have already subsided to a considerable depth.

The three curves representing the more or less pronounced deviations of the climate from normal conditions correspond with the greater or lesser intensity of the processes indicated by the three series of curves on the left. An extreme regressional stage and period of very marked and wide-spread mountain-building will thus correspond with a considerable differentiation of climatic zones and the occurrence of glaciation. An additional cooling power of the ice-sheets causes them to extend very rapidly. This is illustrated by the steepness of the curve on the extreme right. Cosmic factors are now responsible for the division of glaciation into a series of ice-ages. This curve, naturally only represents the glacial and interglacial stages, too, very schematically.

These very abnormal climates have been observed twice since the Cambrian, i.e. during the Upper-Paleozoic and Pleistocene, and it was during these two periods that the differentiation of the flora was so much more marked than during the intervening time.

The same major periodicity of approximately 250 million years is reflected in the tremendous variscian and alpine mountain-building and the formation of basins and throughs. The upper-paleozoic phase of mountain-building lasted for about 50 million years, and the alpine belts began to rise 200 million years later [1]). During the intervening time periods of minor activity occurred, with an average cadence of 50 million years, and were accompanied by lesser manifestations of an abnormal climate as shown in Table II [2]). Our galaxy rotates in about 200 million years. Other figures can be found in geological literature, viz. 300 and 250 million years. This cosmic cycle may thus be said to correspond on the whole with the major periods of terrestrial activity. It is difficult to decide for the moment whether this coincidence is a mere question of chance, but it is obvious that this point will have to be taken into consideration in the further development of science.

D. The above period of abnormal climatological conditions ends even more rapidly than it began. The mountains have been eroded to a considerable degree during the intervening time, and the sea-level has risen slowly but surely.

E. At this stage we find a low relief on the continents, together with vast epicontinental seas and a mild and equable climate extending to high latitudes. A maximum transgression means a minimum difference in the height between the continents and the ocean-floors. Moreover, this stage is pre-eminently a period of tectonic quiet. In the meantime, however, certain belts and basins continue very gradually to subside. Abyssolithic injections of basic melts break through the crust and adjust themselves

---

[1]) These figures agree fairly well with those cited by Kuenen in his article of 1941 on major geological cycles.

[2]) It need perhaps hardly be emphasized that the graphs in Table II should not be considered to be mathematical constructions. The positions of the tops of the curves have only been indicated roughly. Their amplitudes are only of value from a diagrammatic point of view and are unsuited for measurement.

into the steadily accumulating strata of the geosynclines and basins as eruptive and extrusive products.

The chronological correlation of A, B, C, D and E as represented in fig. 94 is wholly diagrammatic. A, B and C combined cover a much shorter lapse of time, however, than the longer stages under D and E.

Our own cycle has already advanced to some extent into stage C.

A fundamental problem

Behind the results arrived at so far there still lurks one essential question, viz. to what deeper impulses must we attribute these phenomena? In what manner might these problematic subcrustal processes produce a periodic recurrence of phenomena as represented schematically in Table II?

We might compare our quest to that of the detective in criminal fiction, who attempts to discover the essential qualities of the mysterious culprit whose traces all lead in one direction. We are no longer content with the classical image of Hades, the old and crippled god of the subterranean realm, for his image is now replaced by a description as found in Table II and explained by fig. 94. We know the culprit's haunts as well as some of his more salient features. Very little has been ascertained as regards his restless infancy, but it is possible to show the main outlines of his cardiogram for the last 500 million years as derived from the movements of the terrestrial crust and the sea-level. His finger-prints are supplied by the magmatic phenomena. What creature is this that breathes so heavily once in every 250 million years, and why does its pulse beat approximately four times during the intervening period? Why are the wrinkles in its face arranged according to the intricate pattern in Plates 1–5? Its skin is not only lined by mountain-belts, but is also pock-marked by numerous basins. Yet nothing is known of the events that brought about these marks during certain specified periods. And there are several hypotheses to explain what was responsible for one of its most striking features — the oceanic depressions.

To return from this geopoetical paraphrase to geological prose: can the subcrustal periodicity be said to be of a physico-chemical nature? Might we attribute it to a periodic system of convection currents, or might it be the result of a combination of these diverse activities? Other processes, too, of which nothing is known at present, may be active in the interior of the earth.

Everyone will have to admit that all views on this subject must necessarily retain a speculative character.

Most readers will be acquainted with the interesting suggestions which Joly and Holmes put forward some years ago, when they attempted to derive periodic phenomena from radioactive processes, and will probably also have read the opinions of such authors as Schwinner, Escher, Holmes, Pekeris and Vening Meinesz on convection currents. Some may have come across Grigg's publication in which this author tries to show that convection currents may in fact occur intermittently, and will have admired the ingenious model by which he sought to demonstrate the activity of cyclic convection currents.

In short, geologists and geophysicists are already striving for a solution, and the hypotheses that have appeared from time to time may perhaps already contain a germ of truth. Nevertheless, the mysterious interior of the earth involves more than just a single periodic process. It is far more complicated than the tentative theories have supposed so far. I hope that the preceding chapters and the accompanying graphs will have made this clear. Geology will have to proceed hand in hand with Geophysics in a combined effort to unravel this most absorbing problem. A solution will only be found when it will be possible to indicate that a logical and necessary correlation of cause and effect exists between certain events in the earth's interior and those diverse phenomena to which I referred briefly in the title as *the Pulse of the Earth.*

References

BARRELL, J. *Rhythms and the measurement of geologic time* (Bull. Geol. Soc. of America, 28, 1917).
BUCHER, W. H. *Deformation of the Earth's Crust* (Bull. Geol. Soc. of America 50, 1939).
DALY, R. A. *Our mobile earth* (1929).
GRIGGS, D. A. *A theory of mountain building* (Americ. Journ. Sci. 237, 1939).
HOLMES, A. *Contribution to the theory of magmatic cycles* (Geol. Magazine 62, 1925; 63, 1926).
JOLY, J. *The theory of thermal cycles* (Gerlands Beitr. Z. Geophysik 19, 1928).
KUENEN, PH. H. *Major geological cycles* (Proc. Nederl. Akad. v. Wetenschappen 44, 1941).
UMBGROVE, J. H. F. *Periodicity in terrestrial processes* (Americ. Journ. Sci. 238, 1940).
VENING MEINESZ, F. A. *Evenwichtsverstoringen in de aarde* ("De Ingenieur" 1940).

APPENDIX

Description of Plates 1–6

The following pages will discuss and enumerate the data on which the maps have been based. This part should only be consulted by those geologists who wish to examine questions of detail. The citations of literature refer to the bibliographies at the end of Chapters II and III.

Plate 1. Caledonian epochs of compression

We will begin with Spitsbergen. The epochs of compression in this part of the world are probably Ardennian, according to Frebold, and possibly also Taconic and Erian. The eastern limit of the geosyncline is situated in North-eastland, where the caledonian strata become less and less disturbed. The geosyncline's northern continuation is a problematical one. It may possibly run via the submarine ridge uniting Spitsbergen and the North of Greenland with the so-called Smith-Sund geosyncline of Northern Greenland. There are no data, however, to prove the exact age of the folding in this area. It might therefore be a caledonian geosyncline, but it might also be a zone of later folding (Teichert 1939, p. 146, 147). A feebly folded caledonian area exists south of Spitsbergen, on Bear-Island (Devonian resting unconformably on Hekla-Hoek formation). A later epoch (Erian) than that observed in the western sector (e.g. the Trondheim area), where the principal phase is the Ardennian, was reported in Scandinavia, around Oslo. Baily asserts (1938, p. 39, 73) that the Ardennian also constitutes the principal epoch of the Highland border-zone of Scotland. Other epochs too, seem to have been observed in England (von Bubnoff 1930, p. 131, 132), the Taconic being particularly evident in South Wales. A section of the Pentland Hills near Edinburgh shows a strong Erian and weak Mid-Devonian phase of movement (fig. 19). The complicated history of the Caledonian zone of England is clearly demonstrated by Owen Th. Jones in his address of the year 1938.

Though the Caledonides in Southern England now lie buried beneath the sediments of a variscian geosyncline, there are still many indications in Europe that Caledonian movements extended over a large portion of the European Continent. The presence of a caledonian unconformity was revealed in many parts of the so-called Saxo-Thuringian zone, forming part of the later, variscian sector of Europe. A taconic unconformity, extending as far as the central mountains of Poland, is clearly visible in the Rheno-Hercynian zone.

Schwinner claimed as a result of microseismic observations that the Scottish Caledonides should not be connected with the Scandinavian, as the former appear to bend around towards Iceland. The caledonian folds of Scandinavia seem to turn in a southerly and south-easterly direction.

It is still doubtful whether a branch of the caledonian geosyncline of Scandinavia and Ṣcotland extended towards Western Europe, as Stille assumed (1924, p. 72). (The massifs of Brabant, Rocroy, Stavelot and Condroz, and the late Caledonian folding in the Boulonnais). This was denied by van Waterschoot van der Gracht (1938, p. 1393), who regarded the Brabant Massif, to mention but one example, as a precambrian area, which was folded in a NS. direction, and the enveloping early paleozoic sediments as strata which were down-folded into the massif during the Variscian epoch. Many others, however, believe that the intensely folded early paleozoic sediments are in fact unconformably covered by upper-paleozoic deposits, and that the plane of unconformity itself again was folded with the later sediments. An example of this is found in the French Ardennes, near Fépin (fig. 18). Nevertheless, the absence of traces of a caledonian unconformity in the Moldanubian zone appears to be indisputable (see under the Variscides). A caledonian unconformity can be found in the south of Spain, and a taconic in Portugal (Born, p. 667). The above shows that the situation is far from being a simple one, and scarcely more comprehensible. For Caledonian movements have at any rate been observed north as well as south of the Moldanubian zone in Europe.

While Caledonian movements, and partly

folding, are thus known to occur on the eastern side of the Atlantic and to cover a large area, all that is found at the other end is a comparatively narrow zone, directed approximately NE.-SW. The latter can be followed from New-Foundland to New-York, and the variscian zone of folding around the coast appears to have been folded at an earlier date, i.e. during the Taconic epoch (cf. Van Waterschoot van der Gracht). Another important point is that the principal epoch of folding in North-America is the Taconic. Later Caledonian phases are not observed in these parts, though the Scottish and Scandinavian Caledonides have often been taken for a continuation of the American belt. It may be that the New Foundland Caledonides continue in the direction of the eastern coast of Greenland. A certain similarity in the marine faunae would seem to prove this. It is as yet impossible, however, to give an accurate description of the Caledonian epochs of folding in this last area (Teichert 1939, p. 150). The caledonian age attributed by Koch and others to a zone of folding in the north of Greenland does not appear to be sufficiently well-founded. The period of folding might be Variscian, or it might even be a more recent one (Teichert, 1939, p. 146, 147). However, even if it were true that a connection existed between the American Caledonides and those in Greenland, this would not preclude the existence of a branch extending towards Europe. Still. as explained above, the area of caledonian folding is a far more extensive one in Europe than in America. And, secondly, the principal epochs of folding of the New-Foundland and the British-Scandinavian Caledonides are different ones. Hence the extremities could not at any rate have been directly connected.

Very little is known of a possible westward continuation of the Caledonides of the Atlantic coastal area of North-America. Taconic movements in large areas in the Rocky Mountains, and locally in the mountains along the coast, are indicated by the absence and feeble development of the Silurian. These indications are lacking in the Canadian Rockies.

As we move south, evidence of Caledonian movements is only found in a few places in the South-American Andes. Gerth (1932, p. 104) mentioned a taconic unconformity in the Argentine Province of San Juan, and also reported the presence of a taconic and later-caledonian unconformity in Peru. Another fragment of caledonian folding may possibly crop out in the Venezuelan Sierra de Merida. Further evidence of such folding may — apart from that observed in the Cordillera — possibly be found in the southern part of the Sierra of Buenos Aires (Gerth 1932, p. 187–190). The Brasilides and Pampine Sierras, which were once regarded as an area of caledonian folding, are clearly of precambrian age (Gerth, 1932, p. 80). The remarkable appearance of folded cambrian quartzite in the north of the Argentine should furthermore be noted. This quartzite is covered uncorformably by ordovician deposits, beginning with a basal conglomerate. The phase of folding is in this case probably Stille's Sardic epoch, which was also observed in Sardinia, the Himalayas, China, Central Asia and New Brunswick (Stille 1939, p. 771).

As very few data are available at present, nothing can be said for the moment of the trend of the early paleozoic mountains, nor can anything be said of an eventual link between these mountains and others situated on other continents. It may be assumed, however, that the Caledonides probably extended (roughly) in the form of an arc, like the Variscides.

Three areas in Africa are known to contain evidence of probable caledonian folding. Hennig (p. 59, 60) reported the presence of caledonian folding in the massif of Ahaggar, but Krenkel has pointed out since then that its caledonian age has been refuted (p. 1442). Hennig also showed that terrestrial deposits, corresponding approximately in age with similar deposits in the Cape System, occur in folded chains in the Congo, and that analogous and intensely folded paleozoic beds have apparently been covered by Permian in the Katanga. He regarded the latter as Variscian chains but Krenkel allows for the possible occurrence of caledonian instead of variscian folding in the "Katangides". In the latter's opinion the "Congolides", (as well, consequently, as the lower Silurian folding („intensive Rahmenfaltung") of the "Griguaides" (p. 684, 912)), are undoubtedly Caledonian. These belts have not been indicated on Pl. 1.

Leuchs observes that part of a caledonian zone of folding can be followed around the precambrian nucleus of Angaraland (1, p. 176). This zone shows later Variscian movements along its eastern margin. Another caledonian area is likewise known to extend along the eastern and north-eastern margin of the precambrian massif part of which subsided into the Kara Sea. A small portion of such a zone — which is presumed to have existed around the massif of Tsuchktschen — may apparently still be found in the New-Siberian Islands. Upper Silurian was folded with the rest in the caledonian zone around the massif of the Kara Sea. The oldest Caledonian epochs observed along the south-western edge of the massif of Angara appear to be of cambrian age (Obrutchew, p. 134–137, and Leuchs 1, p. 153) and to date from the close of the Silurian in the Salair (Leuchs 1, p. 155).

Caledonian folding also played an important part in Manchuria, the Mongolian Altai and the Russian Altai (Leuchs 1, p. 160, 163, 188). The Caledonides of South and South-East Asia probably occupied a considerable expanse, for caledonian unconformities are observed in many mountains in which evidence is found of variscian or younger folding e.g. in the Kirgises (Leuchs 2, p. 29), the Kwenlun (1, p. 33 and 2, p. 189), the Tianchan (2, p. 117), the Arkatagh (2, p. 231) the Nanchan (2, p. 227) and the south-eastern part of the Asiatic continent. Faint Taconic movements were reported by Lee (p. 108) in the whole area north of the Tchingling mountain-chains. Early Caledonian movements, on the other hand, have been observed in the south of China and in Indo China (Lee, p. 115), but these epochs cannot be determined accurately. No Caledonian belt has been observed in the East-Indian Archipelago, though such a mountain-chain is known to exist in Australia.

A geosyncline, extending along the eastern edge of precambrian Western and Central Australia, and apparently coming to an end in this last area, is known to have existed in Australia in former times. Very intensive Taconic and post-Silurian movements (?Erian) occurred, according to Andrews, and a mid-devonian unconformity was also reported in the eastern sector of Victoria. Generally speaking, the movements appear to grow more recent as we move from the west to the east, and the structure of the Caledonian belt seems to become more and more intricate, consisting of consecutively folded geosynclinal zones. Folding in Flinder's Range, Adelaïde, apparently occurred as early as the Mid-Cambrian, and the geosyncline subsequently receded towards Victoria. It remains to be seen whether the same or whether different Cambrian epochs are involved in Asia. On this last continent movements also succeeded Lower-Ordovician time. The youngest Caledonian movement in the eastern part of Victoria occurred towards the close of the Upper-Devonian (Bretonic epoch). These also influenced the earlier folded zones. Andrews still classified these phases as Caledonian epochs (p. 124, 173), since they represent the final epoch of these ranges' history.

Plate 2. Variscian epochs of compression.

The analyses of the Variscides of Western Europe by Kossmat and other German geologists clearly illustrate this area's complicated zonal structure. When considering the huge expanse occupied by variscian folding in Asia, it should be born in mind that these areas are at least quite as complicated from a structural point of view as those of Western Europe, and that they should not be regarded as merely one, vast geosyncline of tremendous width; moreover the different zones were folded in different epochs. In the European Variscides, too, the geosyncline proper is confined to a narrow zone — the so-called Rheno-Hercynian zone (1 on Plate 2). Yet even here three distinct units can be observed. In the southern sector (e.g. the Schiefergebirge of the Rhineland and Brittany) the folding is Bretonic, and north of this sector — in Normandy, Rocroy, Hohe Venn and Siegerland — probably Sudetic. More recent folding is found in the most northern coal-bearing zone, where the principal epoch is probably the Asturian. Kossmat mentions yet another zone: the Saxo-Thuringian, situated south of the Rheno-Hercynian area (2 on Plate 2). Folding in this zone is of Sudetic age. The Moldanubian zone (the Plateau Central, the southern regions of the Vosges, the Black Forest and Bohemia) is found south of the Saxo-Thuringian zone (3 on Plate 2); it consists largely of precambrian rocks. No caledonian unconformity has been observed here, nor is it possible to say whether variscian folding occurred in this sector. There is no doubt, however, that it was influenced to a considerable degree by Paleozoic movements. Intensive faulting occurred, and was probably accompanied by overthrusting of entire blocks over folded areas.

The Sudetic is the principal epoch of movement in the central region of the Variscides. This phase did not affect the outer zone, which was formed at a later date. Folding in this area was probably Asturian. The central part supplied the outer zone with sediments, and the latter rose as a mountain-chain towards the close of the Carboniferous (the Saalian epoch is scarcely noticeable).

How are we to account for the continuation of these belts? Stille and Born were both of the opinion that they bent around within the European shelf and emerged again in the Spanish Meseta. So little is known of the detailed structure of this last area that it is impossible to say whether it contains any evidence of Central European zones. One thing should certainly be noted, however, viz. the structure of the Meseta runs NW.-SE., and the Asturian Mountains actually do bend around, forming the so-called Asturian Knee. Sudetic, Asturian and Saalian epochs are found in Asturia. I am of the opinion that if the Variscian Mountains of Western Europe curve back towards the south of that continent this would tend to show that the Asturian Carboniferous should not be connected with the northern marginal zone of the Rheno-Hercynian belt, but that it should be regarded as a formation similar to the northern paralic zone (but situated south of the Variscides).

The presence of Variscian folding was partly revealed and partly assumed in the Western and Eastern Alps, the Carpathians, the Iberian-Islands and — finally — Corsica, where the most influential epoch appears to have been the Sudetic. The most important phase in Asturia and the Northern marginal deep of the Variscides was the Asturian, and in the Pyrenees, the Donetz basin and the Ural the Saalian.

The variscian area of North-Africa, which was later influenced by alpine folding, was connected with Europe. This also applies to a belt south of this area, extending as far as the Touareq Massif. Folding in these sectors is bretonic in the south, and sudetic further north (Born, p. 794). The folds run, generally speaking, in a NS. direction, bending eastwards at their northern extremity (Hennig, p. 61).

Though it may be correct to assume that the European Variscides form an independent system, this does not necessarily mean that some of the folds did not branch off towards the west continuing on the other side of the ocean. As an example we cite a connection which may have existed between Northern Europe and the Appalachians, and to this might perhaps be added a second link between the southern part of North-Africa and the northern part of South-America.

Before dealing with America, the possible existence of a connection between the European and Asiatic Variscides should be considered. Very little can be said on this subject. Variscian folding seems to have been observed in the Dobrutcha, and is presumed to exist in the basement of the Carpathian Mountains and that of the Pannonian basin (Cornelius, p. 358) Ancient massifs are found in the Balkans (e.g. the massifs of Rhodos, Pelagon and the Cyclades), but nothing is known as regards the age of the folding. It is probably variscian in the massif of Pelagon and the massifs of the Eastern Alps, where Sudetic and Saalian epochs occur according to Stille. Folding was saalian in Euboea (Stille, p. 105). Variscian, and especially saalian unconformities were observed in many parts of Asia (Philippson and Stille, 1924, p. 118). The same epoch also characterizes the Southern Variscides of Asia, and can for instance be observed in the Himalayas and Burma (Stille, p. 119).

Thus only vague indications can be found in the ancient massifs and exposures of the basement-complex of the Alpine Mountains of an eventual previous connection between the European and Asiatic Variscides. The Ural, which is situated on the border of these continents, will be discussed subsequently, together with Asia. It should be noted now, however, that the principal epoch in the Ural, as in the Pyrenees, was the Saalian.

A probable variscian zone, stretching from the Caspian Sea into the Donetz Range, can be traced via a group of smaller exposures which von Bubnoff christened the ,,Ammodetic Mountains". Folding in this zone is not intensive and can be compared to that in the Jura Mountains. The Donetz basin was formed between the Podolian Mass and the Mass of Voronez. This took place in the Upper-Devonian The first epoch occurred towards the end of the Carboniferous and the beginning of Permian time, the principal phase between the Permian and Upper-Triassic (von Bubnoff 1, p. 213). It has apparently not yet been possible to determine the exact age of these epochs. Stille (1934, p. 105, 207, 141) speaks of saalian and upper-cimmerian folding. The principal epochs of compression in Europe would hence have been the Sudetic and Asturian, except in the regions which have just been mentioned. The Sudetic was particularly influential. It will subsequently be seen that this phase also constituted the major epoch in wide-spread areas of Asia. Opinions vary as regards the principal phase of folding in the Timan Ranges (von Bubnoff 1, p. 196).

Two geosynclines can be observed in the North-American Variscides, i.e. the geosyncline of the Wichita and Arbuckle Mountains, with a W.N.W.–E.S.E. strike, and the great Appalachian geosyncline. The first dates from the Cambrian. Intensive folding in the Wichita mountain-chains was Sudetic (Wichita epoch), and subsequently Asturian (Arbuckle epoch). The intervening Ardmore-Anadarko basin was only influenced by the last epoch. This lower-paleozoic geosyncline probably continued in a S.E. direction, and was overthrusted by the mountain-chains of an upper-paleozoic belt, running from the Marathon Mountains to New Foundland via the curve of the Ouachitas and the Appalachians. The epoch of folding in the Marathon Mountains is the Asturian (Arbuckle epoch). A fainter and later stage of folding probably occurred during the Saalian epoch (Appalachian). The Ouachitas, on the other hand, probably overthrust great stretches of the foreland, as well as the chains of the Wichita-Arbuckle geosyncline (a considerable part of which had already been worn down by erosion) during this same Appalachian revolution. The Acadian (Bretonic) phase, too, was of exceptional importance in the Appalachian chains proper.

For detritus (molasse) from the chains which were then just beginning to develop was later gathered into one of the foreland's basins. This basin, into which all the products of erosion were thus assembled, was folded during the so-called Appalachian revolution (Saalian epoch) and simultaneously overthrust by the older folds of the Appalachian System. The Appalachian epoch may therefore be said to have also influenced most of the earlier Appalachians.

Available data on the structure of the North-American and European Variscides clearly show that both parts cannot be joined without assuming that some fragment has disappeared. Meanwhile, the American Variscides indicate in a most decisive manner that epochs of folding may vary within a relatively small superficies. The former existence of a connection between both formations should therefore not be denied *a priori*, though the epochs of folding of the European and American Variscides cannot be said to coincide entirely.

Mention should here be made of Stille's hypothesis that the American Variscides bent around in a southerly direction like those in Europe. The North-American Variscides, however, extend arc-wise from the Marathon Mountains in a westerly and northerly direction, spreading along the boundary of the Canadian Shield, in which last area it becomes difficult to disentangle them from the more recent structures of the Rocky Mountains. An unconformity has been observed in several places between the Paleozoic and Triassic (North and East California Oregon, Ochoco Mountains), while many of the batholithic intrusions appear to date from the Paleozoic (in the Klamath Mountains e.g.). The variscian geosyncline may consequently have continued around the Canadian Shield and joined up with the variscian girdle of Northern Asia. Another question is whether a variscian branch can be said to have extended southwards. This question is even more difficult to answer than the preceding one. Waters and Hedberg (p. 13) wrote the following in this connection: "In the Caribbean region, because of the absence of early Mesozoic sediments, exact dating of the diastrophism which deformed all Paleozoic sediments before Cretaceous time is not possible. In Northern Central America folding and faulting of the Permian sediments accompanied by intrusion occurred before deposition of the overlying Cretaceous. In Jamaica, Hispanolia and Northern South-America, Cretaceous sediments rest unconformably on older rocks, which may be either of early Mesozoic or of Paleozoic age. The extensive deformation in the Caribbean region is commonly correlated with the Appalachian revolution. Whether it coincides with this or occurred somewhat later is questionable." It is therefore not improbable that a communication existed between the Variscides of North and South-America, running through the area which now constitutes the archipelago of Central America.

Later investigations will also have to solve many problems in South-America, but our knowledge of the Variscides in this continent is such, that the general outlines of Variscian chains can be traced from Venezuela to the Argentine along the western edge of the precambrian mass of Guyana and Brasil, where the ranges turn towards the Gondwanides of Buenos Aires and meet the Atlantic. No area of variscian folding has been observed on the South-American Continent south of the Gondwanides. The epochs of folding have not yet been identified in the greater part of the territory occupied by the South-American Variscides. They appear to date mostly from the Upper-Paleozoic. In the ranges of Buenos Aires, an Upper-Carboniferous (Asturian?) epoch and a final phase which Gerth assumed to be a Permian epoch in 1932 (p. 190), but which he described later (1935, p. 249) as a Triassic (or so-called Gondwanide) epoch, is found side by side with probably caledonian folding. This final epoch, though younger than the phases usually observed in Variscian Mountains, thus coincides with the most recent epoch of folding found so far in the South-African Cape Mountains.

In this case the epochs of compression would consequently correspond on both sides of the ocean. However, Krenkel and Gerth (p. 190) declared the structures of the Sierra of Buenos Aires and the Cape Mountains to be fundamentally different ones, though, as seen above, an Upper-Carboniferous and a Triassic epoch are evident in the last area (Krenkel, p. 593). An area of variscian folding is also found in East-Island (Falkland-Islands). The permian layers are still folded steeply in this region and the latter also appears to contain indications of pre-permian (post-devonian) folding. Baker's latest observations show that an extremely faint movement can be followed in those parts of West-Island where a continuation of the folds of East-Island might be expected. Hence the Sierra of Buenos Aires should not at any rate be imagined to be connected with the western part of the Falkland group by means of sharp curves, but should be considered to extend arc-wise into its eastern sector. Another branch would then connect them with Africa. Besides, these connections do not even appear to be necessary. For not only do the Cape Ranges expire near the Atlantic coast (Krenkel p. 613), but one branch turns around in a northerly direction in the Cedar Mountains, running approximately parallel to the West Coast of Africa. Moreover, the actual Cape Mountains would seem to form a mere northern fragment, of a far larger Variscian Mountain System which — together with its connection with another fragment: the Natal Mountains — is believed to have foundered in the ocean south of Africa.

Which epoch of compression should be considered to have constituted the most recent phase of folding in the Cape Mountains and the Gondwanides? Gerth and Du Toit are somewhat vague in this respect. In "Our Wandering Continents" (1937, p. 81), the latter wrote: „several pulses which attained the maximum during the early Triassic and which appear to have corresponded approximately with the Asturian, Saalian and Pfalzian phases in Laurasia", and on page 310: "The Gondwanide foldings about the closing Permian and early Triassic". To this he added in 1939 (p. 505): "Before the close of the Triassic the bulk of the Cape foldings and much of their subsequent erosion took place". Gerth seems to imply that the Lower-Triassic was the youngest epoch in the Sierra of Buenos Aires. I consequently referred to this epoch in 1939 as a separate "Gondwanide phase" of the Triassic (Table 1, Geol. Mag. Vol. 76). Hennig speaks of a Rheto-Liassic epoch (1938, p. 128. 134). In 1924 Stille referred to it as the Lower-Cimmerian phase, since the most recent epoch of folding in South-Africa occurred subsequently to the deposition of the Beaufort beds (Permian and Triassic), but prior to that of the Stormberg beds (Rhetic or Liassic). Stille was probably right, and in this case, too, (cf. the Dobrutcha), the most recent epoch of the variscian sequence probably occurred towards the close of the Triassic. Baker, too, was of the opinion (1937, p. 27) that the youngest epoch observed in the Falkland Islands was "possibly about the close of Triassic time although it may have been pre-Rhetic".

Leuchs assumed that caledonian and variscian zones had formed around the three largest precambrian massifs in the same way as in Europe, where the Baltic-Russian Shield was first partly surrounded by the Caledonides and subsequently by the Variscides. The mountains of Nova Zembla and the Ural constitute the most western Variscides, and as a low area, containing mesozoic and tertiary sediments, extends between the Ural and the Angara Shield (the basin of Western Siberia), Leuchs concluded that its basement had been folded during the Variscian epoch. The arcuate structure of the Variscides in the Kirgise steppes east of Lake Aral would indeed seem to bear this out (Leuchs, 2, fig. 10 and 14).

As the Saalian represents the most recent epoch of folding in the Ural, folding and mountain-building may consequently be expected to have occurred at an earlier date further east as we approach the Angara Shield (the latter is surrounded by the Caledonides). Sudetic folding has in fact been observed in the Russian Altai in the immediate neighbourhood of the Caledonides in the south-west corner of Angara (Leuchs 1, p. 157, 160). Similar folding is found in the Tarbagatai (Leuchs 2, p. 33), and (south of these) in the Dsungarei (2, p. 41). The Tianchan Range, forming the northern boundary of the Tarim basin, was folded during a late Variscian epoch. The same folding is observed between the Nanchan mountain-chains and the massifs of Tarim and Ordos. The principal epoch of folding in the Kwenlun appears to have been the Bretonic. Earlier phases (Sudetic? Appalachian?) seem to have had but little influence (2, p. 190). In the area of the eastern confine of the massif of Angara, folding apparently occurred during a Lower and a Mid-Carboniferous epoch.

Since the massif of the Kara-Sea is known to have first been surrounded by a caledonian and subsequently by a variscian zone, this last belt may reasonably be expected to belong to said massif (and not to that of Angara). In that case the variscian zone would be more recent than the Variscides that formed around the nucleus of Angara. The youngest epoch in Taimyr appears indeed to be the Saalian (Leuchs, p. 110). Similar folding seems to occur in the Ural. Even overthrust-sheets have been observed in these mountains. The Pfalzian phases in these sectors do not appear to be very pronounced.

In the south, variscian folding was even observed in such Alpine chains as the Himalayas. The Variscides bend around the ancient massifs of Ferghana, Tarim and Ordos. Younger Cimmerian movements greatly obscure the Variscian phases along the southern boundary of the shield of Angara (Transbaikalia).

In South-Eastern China the Variscides continue around the massifs of Eastern China and Indo-China. Lee summarized all areas in which sudetic, asturian and appalachian unconformities are known to occur (p. 125, 133, 139 and 149).

The Appalachian phase is believed to have been the closing and principal epoch in the Tsingling (south of the massif of Ordos). There is evidence that the variscian zone probably extended beyond the actual continent, reaching as far as the Riu-Kiu Islands, Japan and Sachalin, but these indications are very vague. A post-paleozoic and pre-eocene unconformity is mentioned by Hanzawa in the Riu-Kiu Islands, and a post-carboniferous though pre-Turonian one has been reported in the eastern part of Sachalin (Leuchs 1, p. 189). The folded strata of the Japanese Variscides include upper-carboniferous sediments (Stille, p. 122).

Little can be said of the former connection between the Variscides of Asia and Europe, for all that is left of this link is now buried beneath the alpine zone of folding of Iran and Asia-Minor. This matter was referred to when we were dealing with Europe.

Another question at this point is whether the Variscides of South-Western Asia had at one time been connected with those in Australia. Nothing definite can be said on this subject, as a supposedly variscian area of folding has so far only been observed in a few regions in the Indian Archipelago, e.g. in Malaya, Borneo and a small number of other islands scattered in between these areas. This question has a direct bearing on the age of the Donau-formation and similar rocks. They were originally defined as jurassic formations by Molengraaff. I considered them to be older than the Upper-Triassic, and believed them to have been folded during the early Cimmerian. Zeylmans van Emmichoven, however, has pointed out since then (in "De Ingenieur in Ned. Indië 1938") that they had probably been folded during a late Variscian epoch, and subsequently refolded during the early Cimmerian phase. A variscian unconformity has also been observed in Sumatra, near Lake Toba. It is not thought unlikely that late Paleozoic movements affected a number of other islands.

The Australian Variscides extend along the coast of Queensland and New-South-Wales. On the whole the epochs become more and more recent as the area extends further east. A post Lower-Carboniferous phase appears to have been met with on the coast of Queensland, between Cape York and Townsville, and an Upper-Paleozoic epoch (which Andrews claimed to be the Appalachian) was observed in the area south of Townsville. David (p. 61) was of the opinion that the first phase corresponded with the Asturian in Europe.

Paleozoic movements have likewise been reported in areas outside the continent. Piroutet and Wilckens mention Permian lying unconformably on paleozoic formations devoid of fossils in New-Caledonia. A Variscian movement of an indefinite age is found in New-Zealand (Stille 1924, p. 123). It had little influence, as will be seen later when dealing with Mesozoic movements.

It may therefore be concluded that large areas of the Variscian Mountains were already in existence during the Upper-Carboniferous, and that a later Saalian phase confines itself to the marginal areas in all continents, lying on the inside in America and on the outside in Europe, Asia and Australia. The Ural, the Taimyr Mountains and the Pyrenees are examples of Saalian mountain-chains. The Pfalzian had very little influence on mountain-building and is scarcely noticeable in the belts in which it occurred. The still more recent early Cimmerian epoch appears to have constituted the last phase in the Cape Mountains and the Sierra of Buenos Aires.

Plate 3. Mesozoic epochs of compression.

Geosynclines which were folded during the Mesozoic characterize the marginal areas of the Pacific.

A typical example of a geosyncline which was folded towards the close of the Jurassic, and which is known to contain much basic volcanic material, is found along the West Coast of the United States, in the Sierra Nevada. The epoch of compression in this area is called the Nevadian by American geologists. It coincides with Stille's late Cimmerian phase. The early Cimmerian epoch probably dit not affect the geosyncline to any important degree, though an unconformity may be observed locally between the Triassic and Jurassic (e.g. in Central Oregon). The rocks were intensely folded and metamorphosed during the period of transition between the Upper-Jurassic and Lower-Cretaceous, and many batholiths penetrated the mountains which were developing at that time. The geosyncline extended northwards from the Sierra Nevada, spreading over the area which the Cascade Mountains, the Coast Ranges and the Alaska Ranges occupy at present. From there it probably continued by way of the Aleutes, and probably joined the extensive area of Cimmerian folding in Asia. Austrian movements also affected the northern sector, particularly Oregon, Washington and British-Columbia, and Laramide movements may be observed still further north (Stille 1936, p. 141).

Little is known at present of the nevadian geosyncline's southward prolongation through Mexico and Baja California, but a nevadian unconformity would appear to exist in these areas (see Stille 1936, p. 148). The same period of folding and intrusion is observed again in Cuba (cf. Waters and Hedberg, p. 16), and is also believed to occur in the northern part of South-America.

Gerth does not consider it improbable that the area which he described as the "Zentralandine Sedimentationsraum" had been folded towards the close of the Jurassic, and is also of the opinion that a second epoch probably occurred towards the beginning of the Upper-Cretaceous (1939, p. 14). This view is not shared by Stille, who — though admitting that very few local nevadian unconformities are found in this area — considers that an important part of the Andes was folded at a later date, i.e. during the so-called Andine or Subhercynian epoch in the Upper-Cretaceous, but pre-Senonian. Stille asserts that the Andes are the only existing folded chains in which subhercynian compression was manifested so strongly and announced the impending publication of an extensive review of South-American epochs of folding [1]). We are thus only able to mention a few of the arguments which this author set forth in a preliminary paper (1936, p. 144, 145). One of these was that the geosynclinal stage was accompanied by volcanism (the tremendously thick basic porphyrite formation, also known as the diabase-melaphyr formation) up to Cretaceous time, and that plutonism occurred in the form of a multitude of granitic intrusions during and after the folding. It will be seen that the mechanism in this case is the same as that in the earlier folded Sierra Nevada of North-America. On the other hand, Stille compared the eastern part of the Andes to the Rocky Mountains in North-America. However, folding, volcanism and plutonism in this sector are of a considerably fainter type than that in the Sierra Nevada. Stille was of the opinion that there might still be evidence in Central America of the existence of a Subhercynian epoch, though no trace of this phase is found in the North-American Cordillera. Conversely, neither an Austrian nor a Nevadian epoch of any importance have so far been observed in South-America. Plate 3 illustrates Stille's view on this subject. A dot-dash line indicates Gerth's "Zentralandine Sedimentationsraum".

Burckhardt, including many others who followed his example, regarded the porphyritic breccias which are distributed on such a large scale over the whole area of the South-American Andes as products of erosion of a land mass situated in the eastern part of the Pacific. Gerth objected that such an assumption would be incompatible with all that is known at present on the subject of these rocks. He attempted to furnish another explanation of the origin of said formations (1935, p. 279–282). Born mentions an additional area — a "Pacific coastal mass" — which is presumed to have foundered during the Lower-Tertiary (1932, p. 842). Part of the Cordillera, consisting of gneiss, granite and crystalline schists and extending along the Peruvian coast would then represent a fragment of this coastal mass.

Another investigator who dealt with these submerged borderlands was Schuchert (1935, p. 636). One of these masses is, for instance, "Choco", which is said to have constituted the western confine of the northern cretaceous geosyncline of Columbia, supplying this trough of sedimentation with products of denudation.

Folded chains which the coast intercepts are likewise found in Peru. The Peruvian coast turns north-west, and the ancient cretaceous lines of the Andes' structure curve around in a similar direction. The coast bends north at a certain point, but the structures continue to extend north-west and are consequently cut off by the coastline. Born outlines a short connection between the abruptly ending belts (1933, p. 399), but Steinmann surmises that this N.W. branch of the Andes might possibly even have extended far into the Pacific. The hypothetical constructions of Steinmann, von Ihering, Koto, Repelin and others were opposed by Schaffer.

The arc of the Southern Antilles, like the area of South-America, had probably already been influenced by Subhercynian folding, but the Laramide phase appears to constitute the principal epoch in Patagonia and further south (Stille, p. 149).

Wilckens, too, who summarized the various geological data of this region (1933), asserted that the Andes continued towards Antarctica via the loop of the Southern Antilles, and based this conclusion on both the sedimentary and eruptive rocks. Nothing definite is known of the subsequent trend of the West Antarctic orogenetic zone. An attempt has been made to trace them via Antarctica to New Zealand with the aid of morphological data. This area covers no less than sixty degree latitude, of which no geological data are known! The reader is referred to the figures in the publications by Born (1932, p. 855, fig. 381) and Taylor (1940, 3. 2, fig. 1 and p. 5, fig. 2). The little that is known at present of the principal epoch of folding in New Zealand seems equally vague. Born regarded it as the fragment of a late cimmerian chain (1932, p. 763). Stille described it as an austrian zone of folding (1924, p. 150). The various and conflicting arguments appearing in geological literature on the above matter caused some doubt to rise within me as regards the right age of the principal epoch, and I consequently turned to Dr. Marshall, the well-known authority on New Zealand geology, requesting him to enlighten me on the subject. Dr. Marshall's honoured and extensive reply is reproduced below. It should first be noted, however, that

[1]) This book (H. Stille, Einführung in den Bau Amerikas, 1940) appeared when the manuscript had gone to press.

early paleozoic strata, resting unconformably on crystalline schists and beginning with ordovician sediments, occur in New Zealand. A second unconformity can be observed between the Matai series, which comprises jurassis strata, "basal Cretaceous" (Benson 1924, 1, p. 128) and the later Oumara series. The senonian fossils closely resemble an analogous fauna in New Caledonia, Grahamland, Antarctica and Chili (cf. Benson, p. 130, and Marshall). Faulting occurred during the Tertiary and Pleistocene, and was locally accompanied by very faint folding. Dr. Marshall wrote me as follows in 1939:

"I certainly think that the pre-Ordovician folding was less important than that of late Mesozoic. This opinion is based on my view that the only folded Paleozoic rocks are in the Northwest and in the extreme Southwest of the South-Island. I regard the schists of Marlborough (Blenheim) en of Otago as altered Triassic greywackes though some geologists think that they may be pre-Ordovician.

"As to the date of what I think is our critical period we have the following guides:

1. The youngest of the folded rocks are (*a*) Nugget point, (*b*) Kawhia. At both of these places *Inoceramus* of a large deeply sulcated type occurs and in the latter ammonites which I have referred to the genera *Phylloceras* and *Aegoceras*.

2. The oldest of the rocks deposited after the critical period of folding are found (*a*) at Amuri Bluff on the east coast of the South Island, (*b*) at Kaipara and Whangoroa at the North Island (*c*) at Wangaloa, close to the mouth of the Clutha river (*d*) at Hampdee just north of Oemaru on the east coast of the South Island.

(*a*) The fossils found here have not been fully described. However, they definitely include saurians with nearly flat vertebrae, *Inoceramus*, *Belemnites* and a few ammonites; they seem to be late Cretaceous.

(*b*) The fossils at Kaipara include a number of ammonites which I have classified (Transact. New-Zealand Institute Vol. 56, p. 226). These ammonites seem to be close to the Trichinopoly fauna of Lower Senonian age in India.

(*c*) Wangaloa. The fossils from here I have described. (Transact. New-Zealand Institute vol. 49, p. 450). A small degenerated *Belemnites* is the only Cephalopod. *Puguellus* and *Perissaloa* suggest a Senonian age.

(*d*) Hampden (Transact. N.Z. Inst. Vol. 51, p. 226). Here there are two Trigonias."

It cannot be decided for the moment whether the Upper-Cimmerian, Subhercynian and Austrian epochs, or whether several of these phases are involved, for the most recently folded strata contain *Aegoceras*, an ammonite suggesting a liassic age, and the oldest strata on top of the unconformity date from the Lower-Senonian. A noteworthy feature in this area is that upper-cretaceous strata (to all appearances Lower-Senonian) rest unconformably in several localities.

If it is correct to suppose, as Benson claimed, that Lower-Cretaceous is still found in the Matai series, this would show that an Austrian or Subhercynian epoch is concerned. Dr. Marshall, however, observed that present data do not for the moment justify such a conclusion. The three phases which we mentioned above have thus been marked with question-marks on the map. Hence New Zealand may at any rate be said to constitute a fragment of a Mesozoic chain.

Along the eastern coast of Australia "immense pressures operated from east-north-east to west-south-west along the present Queensland coast from at least as far south as the mouth of the Brisbane River to beyond Broadsund" (David p. 86, 87). This process of folding (the youngest known so far in Australia) occurred in post-Cenomanian time. David was of the opinion that it might possibly be the Laramide epoch (p. 86), but its age has not yet been determined accurately, and it is therefore still possible that we are concerned in this case with the Subhercynian phase. Both epochs have consequently been indicated with question-marks (see Pl. 3 and Pl. 4). Mention should also be made here of a hiatus between the Triassic and Portlandian in New Caledonia. It may therefore be assumed that this region was influenced by either the early or the late Cimmerian epoch. The last intensive folding in this area, however, was of tertiary age.

Cimmerian mountain-chains occupy large tracts in Asia. The following facts, which appear in a publication by Leuchs, to which have been added additional remarks from Lee, willillustrate this. The triassic and jurassic sediments are particularly thick between the nuclei of Angara and Tschuktschen and display folding of a very pronounced character. Powerfully overthrust-structures of the alpine type are met with especially in the marginal areas. The early Cimmerian epoch, and a late Cimmerian phase, can be observed in the Werchojansk and Tscherski Mountains (Leuchs 1, p. 175, 177). This area continues around the south-eastern corner of the Angara shield and is then connected with China, where the principal epochs of compression in wide-spread areas have also been identified as Cimmerian movements. The marginal chains of the nucleus of Ordos contain folding which is at least variscian, but the final epoch of compression is the late Cimmerian. Folding and overthrusting are even known to occur in the Upper-Cretaceous (Leuchs 2, p. 95).

Laramide and Austrian phases have been observed in the Nanling mountain-belts and all along the coast (Lee, p. 189, 190). Further south the early Cimmerian epoch was particularly pronounced in Indo-China, and can be traced as far as Malaya. The same may probably be said to have constituted the final epoch in Banka, Billiton and West-Borneo. The south of China is the area of the so-called mesocathaysian geosyncline (Lee, p. 212, fig. 61). Another geosynclinal area (this time in Eastern Siberia) was already mentioned previously. Cimmerian movements are even found in the massif of Angara, viz. in the Jennessei-Lena zone, where the sediments of the basin of Wilui were compressed in slightly undulating anticlines and synclines (Leuchs 1, p. 166). Cimmerian unconformities are known to occur in several sectors of the Tethys zone, e.g. in Pamir (Leuchs 2, p. 155, 158), and early and late Cimmerian movements were observed in the Gulf of Karabugas, along the Caspian Sea (2, p. 9, 11). Late cimmerian unconformities are found in a zone extending approximately parallel to the Ural (1, p. 200). Cimmerian folding has also been reported in Iran, Anatolia, Korea (Kobayashi) and Japan (Ozawa).

The Dobrutcha is the only part of Europe where no folding occurred after the existing early and late Cimmerian movements (Born, p. 705). Early cimmerian unconformities have been observed in other places (in the Carpathians and the Alps). A late cimmerian unconformity occurs in the Crimea, Caucasus and Apennines, and an extremely faint one in the Alps. Both epochs have been identified in the Balkans, and it is known at present that these unconformities do not only occur in the region of Tethys, but that the same are found in France, Germany and also in other continents. To these areas should also be added the Gondwanide chains, which were dealt with above under the Variscides. The Cimmerian unconformities occuring in mountain-chains which were folded during the Tertiary will be found in Plate 4.

Plate 4. Cenozoic epochs of compression.

The North-American Continent, besides furnishing a splendid example of mountain-chains that were folded during the late Cimmerian phase (e.g. the Sierra Nevada), also provides a typical example of Laramide folding (e.g. the Rocky Mountains). Many overthrusts of blocks which were not folded to a very intense degree, all of which dip westwards, are met with north of the basin of Wyoming. Vertical movements along a set of normal faults dominate south of this area. The Laramide movements, however, are not confined exclusively to the Rocky Mountains. A great number of synclines and anticlines occur in the less mountainous region east of these belts. Some details are now known of the synclinal depression in for instance Alberta. This depression runs along the eastern part of the Rocky Mountains (Waters and Hedberg, p. 23, fig. 3). Laramide movements, extending as far as and into the Coast Ranges, have also been observed in the western sector. Yet it is at times doubtful whether the movements involved are Laramide or post-Eocene (Pyrenean) epochs.

It is impossible to trace the exact southward course of the laramide belt. Pronounced Laramide movements, accompanied by the formation of granodioritic batholiths, are known to have occurred in Mexico (Schuchert, p. 34, 35; see also Stille 1936, p. 148). The exact age of the folding and intrusions in many parts of the Caribbean area have not yet been determined. The age in Hispaniola, however, and in Jamaica, Cuba, Bonaire and Curaçao appears to be post Upper-Cretaceous and pre Upper-Eocene. Though sedimentation continues without a break from the Cretaceous to the Eocene in the geosynclines of Venezuela, the Andes, the Cordillera Oriental and the Sierra de Perija, a decidedly laramide unconformity has been observed in adjacent areas such as Northern Central Columbia (Waters and Hedberg, p. 28).

Stille mentions cretaceous and tertiary strata resting conformably (and folded locally) in the "Subandine", eastern part of the Andes of Bolivia and Ecuador. The same may be observed in the Cordillera of Venezuela, except in the coastal Caribbean Cordillera. Younger movements occurred in the northern area during Oligocene and Miocene time, and a series of smaller idiogeosynclinal basins (such as that of Maracaibo) were formed subsequently.

Two additional geological papers have appeared on the structure of the Antilles since Rutten's review (1935) and that of Sapper (1937; the papers referred to are those of Senn, 1940, and Weyl, of the same year). Senn — unlike Suess, Rutten and Hess — assumes that a land or shallow water communication, extending from the Greater Antilles and the Curaçao Ridge, formerly connected the coasts on either side of the Atlantic Ocean. This link would have been shaped as a series of geanticlinal ridges in the Upper-Cretaceous, and the ridges would have been influenced by laramide folding. Senn argued that a connection, which assumed the form of "a submarine plutonic arc in the Lesser Antilles", was brought about between the Greater Antilles and the Curaçao Ridges during said period. The Pyrenean epoch of folding would have elevated this arc above sea-level, causing another volcanic arc to be created on its inner side. The previous connection between the Antillean and Mediterranean region would then have been intercepted. It should at any rate be noted that Weyl reports that a series of mountains turn around in San Domingo, coming to an abrupt end on the boundary of the present basin. It is hard to believe that the Greater Antilles would not have been connected with South-America by means of a comparatively ancient arc-shaped structure. The post Turonian (Subhercynian) epoch observed by Rutten and his collaborators in Cuba may possibly bear out this statement. In the meantime, it is still to be doubted whether another old connection had not existed between the Antilles and the Mediterranean. Senn and Gerth both drew attention to this question in 1940. The possibility of transatlantic belts was dealt with in Chapter VI. The data on epochs of folding in Plates 3 and 4 have been taken from publications by Rutten c.s., Sapper, Senn and Weyl.

Younger movements — particularly the Miocene and Pliocene — had more influence in the northern and central part of South-America (e.g. cf. Gerth, 1939, p. 36, 52) than the laramide revolution. Though the latter may possibly have had some significance locally, laramide folding certainly did not constitute the final epoch in these areas (see Gerth, e.g. 1939, p. 42, 53, 58) This phase, however, appears to have been the last one of importance in Patagonia, and also — probably — in the extension towards Antarctica via the Southern Antilles (Stille, p. 149, Gerth 1939, p. 28).

All that is known of the younger epochs of folding in the Cordilleras of North- and South-America can be summed up briefly as follows.

The most western folded chains of North-America are also the most recent ones. The zone which these belts occupy can be followed from the Californian coastal chains to the Klamath Mountains and the coastal chains of Washington Oregon, the Olympic Mountains, Vancouver-Island and other islands along the coast of British-Columbia, the Alexander Archipelago, the St. Elias Mountains (in which the strike may be seen to turn west, south of and parallel to the Aleutes and the Konyage-Islands).

A series of upper-mesozoic and tertiary idiogeosynclinal troughs formed in between these ranges and the Sierra Nevada-Cascade -Alaska mountain-chains. Enormously thick deposits filled most of the troughs. In the Ventura basin the sediments attain a thickness of approximately 20,000 meters; the Pugget Trough contains cretaceous and later deposits up to Oligocene, and the California Valley (which is still in a geosynclinal stage) 10,000 meters of thick deposits ranging from miocene to recent strata.

The age of the unconformities and epochs of folding cannot be determined accurately as a whole. The Miocene and Pliocene epochs in California probably correspond with the Styrian Attic and Rhodanic epochs in Europe and the last intensive phase probably occurred during the Pleistocene. Stille refers to this last epoch as the Pasadenian. A thorough and lucid description of the intricate structural history of California can be found in Reed and Hollister. Present data show that these recent areas of folding and these basins originated in regions that had already been subjected to folding during the Nevadian phase, and that they also reveal the influence of laramide compression.

Movements were observed towards the close of the Eocene in many places in the vicinity of the Caribbean. Waters and Hedberg assert that the Cordillera Oriental, in the northern part of South-America, and the Sierra de Perija had evolved during an Upper-Eocene or pre Upper-Eocene epoch of folding and the contemporaneous elevation of that particular portion of the Andine geosyncline. The broad geosyncline of Venezuela and the Andes, extending as far as Peru, would have undergone many important changes during this period. "The broad Venezuelan-Andean geosyncline was broken up into the smaller, narrower Orinoco, Maracaibo-Falcon, and Magdalena basins by the rise of the Cordillera Oriental and the Venezuelan Andes, and throughout much of Northern South-America oligocene and miocene sediments rest with angular unconformity on eocene strata" (Waters and Hedberg, p. 39). One of Rutten's recent publications gives a clear picture of the extremely complicated history and structure of these areas (Proceed. Kon. Akad. v. Wet. Amsterdam, XLIII, 1940). A Lower-Miocene movement is known to have occurred during the subsequent history of the Orinoco geosyncline, but the most important Tertiary epoch in this area and the regions of the Maracaibo-Falcon and Magdalena basins dates from the Mid-Pliocene (? Rhodanic epoch), "and they were largely raised above sea-level....". "The Bolivar geosyncline, in Western Columbia, was folded and uplifted into mountains" (Waters and Hedberg, p. 41). It was at this time, too, that frequent faulting affected the islands of the Caribbean, and this gave rise to the surmise that Bartlett Trough, Anegada Passage and other exceedingly deep parts had begun to form during the above-mentioned period, which also witnessed the submersion of "Antillia".

Certain movements, consisting principally of a vertical elevation accompanied by a slight warping of the strata or folding of a widely undulating character aré known to have occurred in Central and South-America during the Pleistocene, and to have continued up to the present day. Elsewhere the process of subsidence and sedimentation is still in progress (e.g. in the basin of Maracaibo).

The areas of folding on the other side of the Pacific can be followed from Kamschatka to the East-Indies via Sachalin, Japan, Formosa, the Riu-Kiu Islands and the Philippines, and a tertiary area of folding may be traced eastward from the East-Indies to the Fijis, and westwards to Europe and North-Africa via Burma, the Himalayas, the Transhimalayas, the Karakorum, Pamir, Iran and Asia-Minor. It would be impossible to describe these areas' diverse epochs of folding exhaustively. We will consequently confine ourselves to a few remarks. In the Karakorum and Transhimalayan mountain-chains the final folding is the laramide revolution. (Leuchs 2, p. 244, 247). Laramide folding (though not as a final phase) is also known to occur in many other Cenozoic belts (e.g. in part of the East-Indies, New Caledonia? and the Fijis). A zonal migration of phases of folding was observed in several belts. In 1938 I attempted to indicate the position of these zones in the East-Indies, and to trace their continuation to the Asiatic continent (Burma). An attempt to correlate the tertiary stratigraphy of Asia (as well, consequently, as the Asiatic tertiary epochs of deformation) with those in Europe generally gives rise to serious difficulties, and I have therefore refrained from using the names Stille gave to tertiary unconformities. Three distinct zones may be observed in the western part of the Indian Archipelago. They are (1) the areas of Malaya, the Riouw Archipelago, Banka and Billiton, all of which were subjected to folding during the pre-Tertiary (early Cimmerian); (2) the idio-geosynclinal basins of Atjeh and Southern Sumatra, which were folded towards the close of the Tertiary (Wallachian?); (3) areas in which the most recent process of folding dates from the Miocene, viz. Western Sumatra and the group of islands west of this region (Styrian epoch?), where younger movements reveal very unpronounced folding, and much faulting along the surface. These three zones can be traced over the Asiatic Continent (in Burma, namely, for a further discussion of epochs of compression in the East-Indies see my publication of 1938). In 1934 Chhibber published a geological review of Burma, in which all that is known of the structure of this area is set forth clearly.

Three separate physiographic zones occur in Burma. These zones (which also differ from a geological point of view) can be summed up as follows. (I) The area of the Shan Plateau, extending southwards to Tenasserim. (II) the so-called Central Belt of Burma. (III) The Arakan Yoma, bounded in the west by the basin of Assam.

(I) This first zone is composed of pre-tertiary strongly folded strata, striking NS. The latest folded rocks date from the Cretaceous (p. 2 p. 210; laramide folding?). This area was land during the whole of the Tertiary (p. 110, fig. 16, p. 211). The continuation of this tertiary land is found in Malaya, the Riouw Archipelago Banka and Billiton, and as far as Borneo. Moreover, when discussing the pretertiary history of the East-Indies it was seen that folding occurred during an earlier period (the early Cimmerian) in Malaya, etc., and not during a later (Laramide) period, as in the Shan Plateau. Burma's eastern zone is bounded further west (in the direction of the Central Belt), by a fault appearing in the landscape as a morphologically very fine fault scarp.

(II) The Central Belt is sometimes also referred to as the "basin of the Burmese Gulf" (p. 2). This tertiary basin subsided to a considerable depth, and its sedimentation bore an intensive character. It constitutes a geosyncline of the same type as that of the geosynclinal basins in Sumatra, Northern Java, etc. It should furthermore be noted that folding occurred simultaneously in both areas, i.e. towards the end of the Tertiary and the beginning of Pleistocene time (Wallachian?). Cotter is of the opinion that the present Gulf of Martaban constitutes a remnant of this tertiary geosynclinal area (p. 212). In the same way, the East-Indian basin of Atjeh and the basin of Eastern Java clearly continue below sea-level.

(III) The Arakan Yoma and the Naga and Manipur Hills represent an area of folding in which no paleozoic sediments have been found, though mesozoic formations (triassic and cretaceous) are known to occur in this region. This area was folded towards the end of the Cretaceous (Laramide epoch?) and thus already formed a barrier during the Eocene between the Burmese geosyncline and the Gulf of Assam (p. 3, 5. 210). It persisted as an ever-rising geanticline (p. 212), i.e. an uplift, during the Tertiary, in which three epochs of crustal movements can be observed, one occurring towards the close of the Mesozoic, a second during the Mid-Miocene, and a third during post-Pliocene time (p. 216). The question at present is: which is the corresponding zone in the Indian Archipelago? Chhibber writes in this connection: "....continues southwards through the Andaman and Nicobar-Islands to Sumatra and Java" (p. 3, 5). The western part of Sumatra (the so-called Barisan geanticline), and the series of islands west of this region probably formed a single zone, which ought to be regarded as the extension of the Arakan Yoma. The formation of the two "geanticlines" (Barisan and the series of islands west of Sumatra) and their separation by a fairly deep deep-sea basin would then date from a very recent post-Pliocene period, and have accompanied the youngest movements of upheaval in the Arakan Yoma. This suggestion probably contains a great deal of truth, for, as stated above, the Tertiary epochs of folding concur in both Sumatra and the islands west of it. Chhibber describes five different zones of volcanic rocks in Burma. The volcanic strip in Central Burma, extending to Sabang by way of the island of Narcondam and Barren-Island and from there to the Lesser Sunda-Islands via Sumatra and Java is specially noteworthy.

These East-Indian zones cannot be followed to western Asia and Europe beyond Burma, for there are too few data to guide us on this subject at present. Nevertheless, the little that is known of this matter gives us good reason to hope that the same will be possible in the future. For several epochs of folding which are absent in the entire chain, occurring in one special zone only, have already been observed in the Himalayas and Karakorum. This was made clear by de Terra (1936). This author states that after the intensive folding towards the end of the Cretaceous ,,the northern sector apparently continued to be land, the southern or Himalayan part resumed its geosynclinal evolution" (p. 865). Very pronounced tertiary folding affected what remained of the geosyncline in post-Mid-Eocene time, but before the Lower-Miocene, and movements may also have occurred simultaneously with the miocene folding in the East-Indies, for de Terra goes on to say: "This orogeny may have continued to the end of the Burdigalian epoch (Lower-Miocene), or at least it may have been locally revived at that time...". After said folding, which may be said to be "possibly subdivided into an Oligocene and a Lower Miocene subphase", the whole area of the Himalayas remained above sea-level. A marginal deep consequently formed in the Mid-Tertiary along the southern margin of the Himalayas, and the Siwalik-layers, which attained a thickness of many thousands of meters (these layers constitute the erosion products of the Himalayas and might be compared to the molasse of the Alps) were deposited within the trough. The layers were folded during the Pleistocene. The southern chains of the Himalayas were overthrust to the south, and "the Karakorum and adjoining regions suffered a broader uplift, during this process" (p. 867). The chronology of the various epochs, together with the fact that the latter only occur in particular areas, remind one of conditions in the East-Indies. De Terra, on the other hand, drew attention to certain similarities between these regions and the Alps (p. 868).

A zonal succession of crustal movement is again observed in Iran, where the chains emerge from the narrow strip of strongly compressed mountains along the precambrian massif of India, attaining a freer development in this region. The youngest movements in Iran have been reported in the outer or most southern chains. De Böckh speaks of a Wallachian phase (p. 155 in Gregory 1929),and also indicates the presence of Laramide, Pyrenean and Savian epochs (p. 156). The most recent formations in this sector consist of the Mesopotamian marginal deep and the Persian Gulf. The latter may be compared to the Siwalik trough south of the Himalayas. Former authors mention a precambrian nucleus in Persia, which was supposedly surrounded by folded chains. De Böckh draws a parallel between this "nucleus" and the Pannonian Basin in Europe. Baier's latest statements, however, show that there can be no question of the existence of such a massif. The idea of an Iranian "nucleus", or of the existence of the massifs of Gobi and Tibet, to which many authors formerly adhered, should therefore be abandoned, as the same is incompatible with present data. In 1939 Arni observed the following five structural units in the westward extension (Anatolia). From north to south we find: the marginal folds of Anatolia-Iran, the Iranides, Taurides, Anatolides and Pontides. The youngest epochs of compression, i.e. the post Miocene (Attic?) and post Pliocene (Wallachian?) appear again in the southern area, while a hiatus characterizes all zones in the Oligocene (it is difficult to determine whether we are concerned with a Pyrenean epoch in this case, or a Savian). Other (Laramide and late Cimmerian) phases, too, are known to occur in the Northern Anatolides and the Pontides. The massifs of Asia-Minor probably affected the trend of the Tertiary chains. Leuchs (1938) assumed the existence of three separate nuclei, and referred to them as the Karic-Lydian, Lykaonian and Halys Massifs.

The older phases also had considerable influence on the formation of chains that branch off towards the Caspian Sea north of Persia. The Oligocene Savian epoch seems to have been responsible for the most intensive and also for the final process of folding in this area, while early cimmerian, late cimmerian, austrian and laramide unconformities, too, have been observed in this sector (Leuchs 2, p. 9, 11). Von Stahl asserts that the principal epochs of folding in the Caucasus are post Cretaceous (Laramide) and post Miocene (Attic?). A miocene (post burdigalian) unconformity has also been reported in this region.

The Caucasus ends on the shores of the Black Sea. The Dobrutcha is the only range that can be regarded as its continuation west of the Black Sea via the Crimea. However, tertiary formations appear to lie quite undisturbed in this area (Born, p. 705). This region's most recent epoch appears to be the late Cimmerian.

I will deal briefly with the extension of the Tertiary chains of Asia-Minor in the direction of Europe, for the Alpine chains in this last continent (especially the Balkan Mountains, the Carpathians and the Apennines) still raise many problems which can only be solved by an extensive study of the various details. Moreover, a complete review of data on this subject can be found in one of Born's publications. The Karic-Lydian Massif may possibly continue into the nucleus of the Cyclades in the Aegean sea. The pre-mesozoic massifs of Pelagon and Rhodope are found in the Balkans. The Tertiary folded chains formed in the intervening area and around the massifs in the same way as the Carpathians settled around the Pannonian Massif. The latter has since subsided and now constitutes a basin filled with late tertiary deposits. Folding in this sector was affected by neighbouring Alpine movements and is not very noticeable. It seems highly probable that another massif, which lay above sea-level during the Pliocene and supplied the geosynclinal area of the Apennines with erosion products, foundered in the Tyrrhenian Sea (Born, p. 714).

It is exceedingly difficult to reconstruct the original connection between fragments which are found to-day in the surrounding areas of the Mediterranean. Their problematical character may perhaps best be illustrated by the many conflicting hypothetical constructions which are met with in geological litterature (see e.g. Born, 1932, p. 722, fig. 284). All the epochs on the map have been adapted from Stille (1924).

The tertiary area of folding in Spitsbergen (Frebold) is not of a deep-rooted alpine type.

Our next step will be to examine the northward and eastward continuation of the East-Indian alpine zones. It cannot be denied, however, that our inadequate knowledge of the tertiary and pre-tertiary history of the Philippines makes it very difficult to follow them north. Much more is known of Taiwan (Formosa). This island's backbone consists of the so-called "Slate-formation", in which metamorphosed lower-tertiary deposits were observed. These contain *Assilina*, *Discocyclina* and *Camerina*. It should be noted that a conformable sequence of Miocene and Pliocene is also apparent in this island. Yabe and Hanzawa reported the existence of foraminifera suggesting a tertiary-f age in the lower strata of this series (*Nephrolepidina* was found together with *Miogypsina*, but no *Eulepidinae* were observed).

The exact correlation of the "Slate-formation" and the younger tertiary series cannot be defined at present, but it is generally believed that they are separated by a hiatus. Yabe and Hanzawa expressed a similar opinion when they wrote that ".... in the Neogene time (excluding the very early part), an island of Eocene rocks, intensely folded and variously metamorphosed, came to existence at the present site of the backbone range, the island was surrounded by the Neogene seas in which the sediments of the Kaizan-, Byôritsu- and Shokkôzan beds were deposited in upward succession".

This tertiary folding must therefore have succeeded the Lower-Eocene and preceded the "later part of the Miocene" (Tertiary-f). We find it impossible to indicate it more accurately. The history of the Riu-Kiu Islands corresponds in some respects with that of Taiwan. It is marked by the same period of post Eocene regression and folding. Hanzawa assumed that a Stampian or Aquitanian epoch was involved in these islands, as in Taiwan (Hanzawa 1935), and that the same "Burdigalian" transgression coinciding approximately with Mid Tertiary-f or Bebuluh transgression affected both regions.

The tertiary areas of compression clearly continue eastwards in New Guinea, though it is impossible to outline their exact course and extent, as there are too few data on this subject. There can be no doubt that the New Hebrides and the remaining fragment, situated towards the extreme east — the Fiji-Islands — form an integral part of said continuation. The geological survey which Ladd published in 1934 concentrated particularly on the most southern of the two larger Fiji-Islands, i.e. on Viti Levu. The geological history of this island contains interesting points of analogy when compared to that of the East-Indian Aechipelago, e.g. an upper tertiary-e (or Bebuluh) transgression, movements in the lower part of Tertiary-f plus g, which a former author — Brock — regarded as an important epoch of compression, but which Ladd preferred to classify as local disturbances (important movements in the East-Indian Tertiary-f). Other examples are: (1) the thick deposits of marl of the Suva formation, containing many molluscs and foraminifera (dito in e.g. North Java and the eastern part of Sumatra) (2) the occurrence of important faulting towards the end of the Tertiary, which is generally supposed to have accompanied the foundering of large parts of Melanesia (the same phenomenon was observed in many islands in the Indian Archipelago, and is in this case regarded as a contemporary process of the formation of the deep-sea basins), and (3) intensive pre-tertiary folding followed by a subsequent period of erosion, i.e. the presence of land which was first flooded by a shallow sea in Neogene time. This flooding occurred at an earlier date in other regions, e.g. in the area which is now occupied by Eua in the Tonga group (the oldest tertiary formations in the East-Indies rest unconformably over this whole area; they are of eocene age in some sectors, but neogene elsewhere, e.g. in the greater part of Sumatra).

All that is known at present of the New Hebrides has come down to us from earlier investigations. Mawson reported that "extensive submarine beds were accumulating above the folded Miocene series" (l.c., p. 471). Chapman's research has shown that these miocene strata belong to the *Lepidocyclina*-bearing tertiary deposits. According to Mawson the folds point in the direction of the New Caledonian ,,foreland". A series of tuffs and marls several thousand meters thick, which probably cover the youngest miocene and pliocene rocks, appear to rest unconformably on the folded miocene strata. Above this sequence we find coralligenous limestone which in some places has been elevated approx. 700 meters above sea-level.

Miocene folding in the New Hebrides may possible have occurred simultaneously with the movements in Viti Levu. Remarkable enough, there appears to be good reason to suppose that this same epoch of folding is also present in New Britain (= N. Pommeren). For "steeply dipping older Miocene sediments overlain unconformably by a gently-folded Pliocene series of foraminiferal sediments and tuffs" have been observed in this island, and older rocks were found in addition to these formations.

I still agree with Mawson that "if at any time the New-Hebrides ridge formed continuous land connected in the North or elsewhere with other land-masses, these conditions are most likely to have prevailed in the early history" (Mawson, 1905).

It seems improbable that these late tertiary areas of folding would have continued towards New-Caledonia and New Zealand, for the last intensive compression occured at an earlier date in New Caledonia, i.e. towards the close of Eocene time (this area has since continued above sea-level, according to Piroutet), and the principal epoch of folding is also a far older one in New Zealand, as shown above. A second epoch of compression (the Cimmerian) succeeded the Triassic, preceding the Portlandian transgression. A third (Laramide) succeeded the Cretaceous and preceded the Lutetian transgression. A fourth and final phase, which had considerable influence, occurred after Eocene time (these facts are from Piroutet; see O. Wilckens 1925).

Plate 5. Chronological analysis of the continents.

The facts laid down in Plates 1–4 have been combined in Plate 5 so as to form a single picture. In those cases in which regions were subjected more than once to geosynclinal subsidence or folding, only the most recent epochs of compression have been indicated in each specific continental zone. Older movements and matters of detail will be found in Plates 1–4.

The general aspects of continental structure were outlined in Chapter II. Caledonian and later zones were dealt with in the descriptions of Plates 1–4. The vast precambrian areas depicted on Plate 5 consequently remain to be discussed.

A division of the precambrian foundation of the continents was recently made possible by the determination of the absolute age of several rocks and the result is that we are now able to distinguish between at least some groups of folded belts. This subject was reviewed by A. Holmes in "The Age of the Earth". Yet in spite of the promising nature of these first results, it is not yet possible to write a more or less connected history of the precambrian areas. In dealing with the structural history of the continents, we began, therefore, with the Paleozoic and for the moment merely wish to draw attention to the huge expanse which the precambrian areas occupy. The latter are not only known to extend over the greater part of the continents (the Baltic-Russian Shield, Canadian Shield, Africa, Arabia and the largest part of Australia, Antarctica and South-America), but fragments are also observed in mountain-chains that were subjected to folding during more recent periods. Extensive areas are also known to lie beneath the sea.

The precambrian basement occurs in the intervening area between the Cordilleras of West-America (British Columbia, Colorado, etc.), and locally in the folded mountain-chains themselves, e.g. in the Bighorn Mountains and the Black Hills of the Rocky Mountains geosyncline. Precambrian rocks are also found in the more recent European belts, large fragments of which may for instance be observed in the Moldanubian zone of the Variscides. In the Scottish Highlands precambrian rocks with a normal sequence of cambrian strata may be seen plunging beneath the caledonian overthrusts. Precambrian massifs presumably also exist in the North-American Variscides.

In South-America, precambrian formations appear in the Pacific coastal areas of Chili and Peru, and are also met with between Trinidad and the most southern point of the continent in the Cordillera. The former expanse of precambrian formations in South-America was not only a far greater one than that of the present continent, but it also had a totally different shape (Gerth 1932, p. 80, 81). Precambrian rocks are also found protruding here and there from the mountain-chains and younger deposits in the basins of Australia and Africa.

A number of shields, situated between folded mountain-chains that formed around them subsequently, exist in Asia, as in Europe (two large shields and a few smaller nuclei are found on the European Continent, viz. the Baltic Russian Shield, stretching southwards to the massif of Ust-urt (Leuchs 2, p. 12), and the Barentsea Massif with its south-eastern spur: the massif of Putkow Kamen). East of the Variscides of Nova Zembla lies the Karasea Massif. Its precambrian basement may still be observed in Taimyr and a few islands. This massif, however, is covered for a large part by a shallow sea. The same applies to the massif of Tschuktchen (Leuchs 1, p. 180). Precambrian rocks from the shield of Angara (Asia's central nucleus) crop out in the Anabar, Baikal, Aldan and Jennessei horsts, while basin-shaped depressions characterize other parts of this shield. These depressions were filled with later sediments (e.g. the graben of Tundra and Lena, and the basins of Tungus and Wilui). The massifs in Eastern and Southern China, too, occupy a considerable expanse. It may probably be assumed that the basins of Tarim and Ferghana originally constituted nuclei, situated at a greater height and subsequently surrounded by the Variscian folded chains. These nuclei have since subsided and been covered by a thick layer of sediments. Mention should also be made of the precambrian mass of India (and Ceylon), which, unlike the areas previously cited, only played a passive role in the geological history of Asia (this had already been shown by Leuchs). Precambrian rocks are also found in the area between the younger formations — in the Paleozoic and later mountain-chains extending among these massifs (e.g. in the Kwenlun, Karakorum and Himalayas; see Leuchs 2, p. 189, 243 and 258).

The Barentsea Massif is at present a shelf and lies at a depth of between 200 and 400 m. below sea-level. The course of a few rivers can still be traced in this shelf's relief. This area (Frebold, Spitsbergen p. 172, fig. 81), together with that of the Karasea Massif (Frebold, p. 171, see also Holtedahl's bathymetrical chart for the northern seas) lay above sea-level in comparatively recent times. In this last case, it is also possible, when east of Nova Zembla, to reconstruct the course of a few rivers in the submarine relief of the floor. The following indications lead Frebold to conclude that the Barentsea Massif was probably part of a precambrian area. In the eastern portion of North-East land the dislocation of the rocks of the Hekla-Hoek series gradually diminishes. The earlier movements and Variscian and Tertiary epochs only occur in the western part of Spitsbergen. The eastern sector has not been influenced by said movements. And finally, it should be noted that precambrian formations crop out along the coast of Putkow Kamen.

Plate 6. Basins of the second group.

Chapter III reviewed the different types of basins, but only a few examples of the second group were included in the discussion. These will now be dealt with in the order in which the basins have been numbered on Plate 6. The figures inserted between brackets refer to the corresponding numbers of the basins. Attention will also be paid to a few submarine basins towards the end of the Appendix.

It is not always easy to determine the chronological correlation of the formation of discordant basins and certain internal processes, for all kinds of difficulties arise when we attempt to do so. One is that few data are known of the basins' stratigraphy and structures except in a few cases, and a second that their contents have in many instances only been folded very faintly and at times even to an almost imperceptible degree. We are consequently faced with the question whether the oldest sediments cropping out along the edge of said basins ought in fact to be regarded as the oldest depositary strata, as it might be that older deposits, which settled during the first subsiding movement, have since been covered unconformably and therefore hidden from sight by more recent layers. As subsidence is known to have been repeated in some basins after a pause in the first downward movement, the possibility of its occurrence in other less well-known basins should not be discarded altogether. The result is that the beginning of subsidence cannot be determined accurately in all cases. Still, it is apparently possible to indicate the large outlines of the beginning of subsidence in spite of these difficulties. Some may feel that the notations of a few basins on Plate 6 need altering. To mention one example — the Australian North and Desert Basins may have begun to subside during the Devonian instead of the Carboniferous and the beginning of the subsiding movement would then have to correspond with a late Caledonian instead of an early Variscian epoch. The history of the basins of Western Germany and the North Sea probably began in Triassic (Rhetic) time, and some might perhaps wish to connect it as such especially with the early Cimmerian phase. Yet marine sedimentation and that of the movement of the adjacent positive elements commenced as early as the Permian, and this may explain why I chose to connect their origin with a late Variscian (Saalian) epoch. A quantity of other examples might be given, but to my mind the existing facts — though very incomplete and presenting many difficulties — nevertheless corroborate the conclusions arrived at in Chapter III.

We will deal with North-America first. The Canadian Shield contains a number of more or less extensive areas covered by paleozoic strata. These should be considered to represent remnants of a veneer which was originally still much more extensive. That such remnants should be found in those areas in which they actually occur may be due either to a basin-shaped depression or to later down-folding or foundering along faults. Morley Wilson (in Ruedemann, 1939, p. 234) supposes that either the first or the last possibility, or a combination of both, affected the neighborhood of Hudson Bay and Southampton-Island during the Mid-Ordovician, Upper-Ordovician, Silurian and Devonian, and expresses the view that all three may have jointly been responsible for the preservation of the paleozoic sediments in Ottowa Valley. The central basin of Hudson Bay (no. 1) reminds one very strongly of an original depression.

The foreland of the Appalachians and the Ouachitas consists of the "interior lowlands". The basement is composed of the prolongation of the precambrian Canadian Shield, and is buried under paleozoic and younger strata. The precambrian foundation has a very irregular surface and is marked by many basins and domes or "uplifts". In the neighboring area of the Variscides the most prominent examples of the positive elements are the Adirondack and the Cincinnati, Nashville and Ozark domes (see Plate 5). For further details the reader is referred to the tectonic maps of King and Van Waterschoot van der Gracht. These include a clear survey of the intervening negative elements such as the basin of Pennsylvania (no. 5), which is directly opposite the Appalachians, containing a maximum of 6,000 feet of carboniferous and permian strata. Other basins are those of Michigan (no. 4), Illinois (no. 3), the arge mid-continental basins (no. 2), etc. The anisochronous frame of the permian Texas basin (no. 6) was mentioned in Chapter III.

As mentioned above, the Marathon-Ouachita Mountains probably form the frontal chains of a large Variscian belt which was buried under recent sediments of the Gulf Coast. These deposits assumed a semi-circular basin-shaped structure (no. 7) and include a sequence which originated in the Lower-Cretaceous after the Nevadian epoch of folding. It continued up to the Pleistocene with many a hiatus, the most important being the laramide unconformity. The thickness of these strata, which continue under the sea as far as the edge of the shelf varies between 500 and 8,000 m (Stephenson 1939, p. 530, Pl. 1). While sediments were thus elevated above the sea-level in the Gulf Coast area and faintly folded and intersected by numerous faults, the central part subsided in the recent geological past, ultimately forming the Gulf of Mexico. The Gulf Coast basin and Gulf of Mexico probably originated on a variscian basement and should therefore be classified as formations of Type IV. The Gulf Coast basin was formed after the Nevadian phase of folding, and its subsiding tendency was resumed fairly recently (Gulf of Mexico).

Two large basins occur in the precambrian block of South-America, viz. the Amazone and Parana basins (no. 8 and 9). The first originated after the process of taconic folding (Gerth, 1, p. 96), and includes unfolded Gotlandian up to Carboniferous. The basin of Parana, however (see Gerth, 1932, 1, p. 144), was formed after a late Caledonian epoch. It was filled with devonian strata; carboniferous deposits are absent. Its subsidence was resumed towards the close of the Paleozoic, resulting in the deposition of upper-permian and triassic strata, and this was succeeded by large outflows of lava. The sediments have not been affected by folding and total approximately 2,000 m.

The most northern European basin, i.e. that of the Barentsea (no. 10), has an anisochronous frame, as shown in Chapter III.

The area of subsidence in the north-western part of Europe extends EW. towards England from the Russian Shield, and NS. to Belgium from Denmark. This area extends along the edge of the massif of Brabant in Belgium. Three separate basins are observed here, each one having a more or less distinct history. Ridge-shaped elevations of the paleozoic basement with a WNW.-ESE. trend emerge in the intervening areas. N.-S. and NNE.-SSW. trends, too, are found in the later mesozoic and younger history as faintly undulating structures, faults and graben. The horst and rift-forming movements of blocks in the foundation were also accompanied by simultaneous folding of the sedimentary layers with a general WNW.-ESE. trend (the so-called Saxonian folding in the Jurassic, Cretaceous and Tertiary). Typical examples are the axes of the Teutoburger Forest and of Egge-Osning, the anticline of Bray, and that of the Artois-Boulonnais in the basin of Paris. The prolongation of the last anticline is found in that of the Weald (S. England).

In North-Western Europe, subsidence is known to have begun in the Upper-Permian (Zechstein), but the division into separate basins dates from the Upper-Triassic. The basin of North-East Germany is since then separated from that of North-West Germany by the so-called axis of the Elbe, or Pompeckji axis. The geanticline of Erkelenz is in turn situated between the basin of North-West Germany and the most western tertiary basin of the North Sea. Its most western boundary is formed by the Pennine axis in England.

The basin of North-West Germany (no. 12), which is described by Stille as the "Niedersächsische", originated in Rhetic time and contains an almost complete sequence of marine Jurassic and Cretaceous. In Northern Hannover the sediments amount to some 6,000 m excluding permian and triassic strata (the triassic total at least 1,300 m, while the upper-cretaceous in the basin of Munster — which was formed during the late Cimmerian epoch — attains a maximum thickness of more than 1,400 m.). The southern margin along the edge of the "Schiefergebirge" of the Rhineland and Harz and "Flechtinger Höhenzug" consists of a series of faults.

The sediments are considerably thinner in the basin of Eastern Germany (this is not indicated separately on Plate 6). Moreover, marine jurassic and cretaceous strata are absent. This lead Van Waterschoot van der Gracht to doubt whether it were possible to speak of a basin in this case (1938, p. 1379).

The basin of the North Sea (no. 11) was filled by a sedimentary sequence ranging from zechstein deposits to pleistocene. The older formations are only known from deep-borings along the edge. A fine illustration of the subsidence during the Tertiary and Pleistocene is found in a map of Van Waterschoot van der Gracht of 1938.

The south-western corner constitutes the basin of London, but the latter only originated as such from post Upper-Cretaceous time (Laramide epoch) onward. It contains a comparatively thin veneer of sediments (approx. 500 m).

In the opinion of Van Waterschoot van der Gracht the carboniferous sediments of the north-western sector total about 7,500 m, or possibly even 9,000 m.

The phases of movement observed in the above basins are chronologically related to those appearing in the folded mountain-chains. The subdivision of the large north-western European area into three basins, each separated from the other by SSW.-NNE. geanticlines, was due to the early Cimmerian epoch. The late Cimmerian phase (which Stille subdivides into three further phases: the Dilster, Osterwald and Hils epochs) was responsible for the formation of the WNW. anticlines of Germany and England and the basin of Paris. The paleozoic structure of the basement may be said to be reflected in these "folds"; see fig. 32 (in an extensive publication, Stevens showed that the present geomorphology of Belgium is still influenced by the structure of the foundation). The horst-shaped elevations of Harz, Osning, etc. were formed during the later Subhercynian epoch. (Stille divided the latter into the Ilseder and Wernigerode phase). It was this Subhercynian epoch that raised the saltdomes in the basin of North-Western Germany. The Laramide epoch then caused many new movements to occur along the existing fault-lines and also produced many new faults (the same holds good for the basin of Paris and the district of Mons, in South-Belgium).

It should finally be noted that Tertiary movements are responsible for various unconformities and for the large rift-valleys of the Rhine, the Rhone, the Limagne and the Central graben of the Netherlands. These orginated largely in the Oligocene (Pyrenean, Savian? epochs). Important movements occurred along these fault-systems and the newly formed faults at the end of the Upper-Miocene, (Attic epoch?) and at the end of the Pliocene (Wallachian epoch?). In an area like tbat of Southern Limburg, where movements of a group of horsts and graben occurred along N.W. faults as early as the Mesozoic, Tertiary movements generally took place in the opposite direction.

Another group of large rift-valleys is found on the other side of the Alpine chains, viz. the huge fault- and rift-systems of Africa, the Red Sea and Western Arabia (fig. 58). Both are accompanied by volcanism. Reference was made to these graben in Chapter IV while dealing with magmatic cycles.

The above broad outlines of the general principles of the formation and deformation of the North-European basins will now be followed by a brief summary of the remaining basins.

No. 13 — the basin of Paris — was discussed as an example of a discordant basin in Chapter III. No. 19, on the other hand — the Pannonian basin — was discussed as a type of nuclear basin.

No. 14 is one very similar to the basin of Paris and known as the basin ot Aquitania. This area also shows the influence of numerous phases of movement corresponding with epochs of compression in the Pyrenees and the Alps.

The basins of the Iberian Peninsula (the Ebro-, no. 15; Duro-, no. 16 and Tajo basins no. 17) are considerably younger (Born p. 662, fig. 249). The Ebro Basin originated during the Laramide phase and was filled with tertiary sediments, beginning with eocene strata, while the Castillian basins of the Variscian Meseta were filled chiefly with oligocene and miocene deposits.

Five basins may also be observed in the area of the Baltic-Russian Shield. The precambrian basement crops out in Fennoscandia and the massifs of Podolia and Woronesh. This basement is covered by a relatively thin layer of sediments in the massif of the White Sea and the East-Baltic Ufa and Ust-urt Massifs. The Polish (no. 18), Moscow (no. 20), East-Russian (no. 21) and South-Russian basins (no. 22), together with the area of subsidence of the Caspian Sea (no. 23), originated in between these positive elements. The first three are the oldest. They were filled with paleozoic sediments; these amount to a maximum of 850 m in the basin of Moscow (devonian up to permian strata) and a maximum of 1,000 m in the East-Russian basin (Carboniferous up to Permian). The paleozoic sediments are covered unconformably by mesozoic strata (v. Bubnoff, Europa 1, p. 194 and 203). These basins' contents were affected by feeble folding, resulting in weakly undulating anticlines with an approximate NS. trend coinciding in origin with the epochs of folding in the Ural. These same anticlines, however, were also influenced by more recent movements.

The Polish basin (no. 18) formed in the Caledonian period according to Von Bubnoff (Europa 1, p. 197, 198, fig. 47 and 48), but the subsiding tendency migrated in a SSW. direction during the post-Jurassic. The upper-cretaceous sediments rest unconformably on the older sequence and the younger sedimentation consists of oligocene and miocene deposits.

The depression of the Caspian Sea (no. 23) forms a southward prolongation of the East-Russian basin (no. 21), as it were, and contains sediments ranging from permian to upper-tertiary strata totalling some 2,400 m (v. Bubnoff Europa 1, p. 194). Faint folding occurred in the Upper-Triassic, Lower-Turonian and Lower-Senonian. Its subsidence must have been particularly intensive during the Maastrichtian, and the bottom of the depression, as shown by the deposits of cocolith-ooze, must have lain more than 1,000 m below sea-level (fig. 37). Where the depression intersects the Tertiary folded chains, there are clear indications that the movement was continued until the late Cenozoic. The present shape of the Caspian Sea might be compared to that of a double basin, of which at least the northern half may be said to date only from the Pleistocene (Leuchs 2, p. 16, 23).

On one side the Caucasian Mountains come to an abrupt end along the Caspian Sea, and these chains are intersected on the other side by the coast of the Black Sea (no. 24). I have found very little as regards this basin's history in geological literature, but there is no doubt that it should be classified as a basin of Type IV, for it intersects the chains of the Crimea on the northern side and the mountains of the Dobrutcha at the western extremity.

Von Bubnoff is of the opinion that the South-Russian basin (no. 22) north of the Donetz Ranges originated in the Lower-Liassic (Europa 1, p. 213). Nevertheless, its subsidence is known to have been of an intensive nature in the Senonian, Upper-Eocene and Oligocene. Several Mesozoic phases are observed in its history, one of the strongest being the Laramide (ibidem, p. 214). This same investigator writes (p. 208) that "Man hat den Eindruck als hätte sich nach Auffaltung des Donetzgebirges die Südrussische Senke gleichsam als Fortsetzung und Kompensation für den unterbrochenen Senkungsvorgang ausgebildet". This basin might indeed have to be regarded as a marginal deep of the Donetz Ranges, but its shape and extension over the north-western part of the Donetz Ranges, together with the absence of such a fore-deep along the Ammodetic chains, would render this classification a somewhat doubtful one.

The formation of the basins of Europe shows no direct correlation with the areas of folding in which they are situated. The basin of Paris extends inside the variscian zone. Outside the latter lies the North Sea basin, the foundation of which consists of caledonian and older zones. Both have sunk transversally into existing folded structures of different periods. They are not confined to any given zone of folding, and the time of the initial formation differs but slightly in each case.

Attention will now be paid to the basins in other continents, particularly to their position and the time of their origin. We will first take a few examples in Africa. Krenkel (p. 560) writes that the site of the Karroo basin was already occupied by a depression before the beginning of sedimentation in the Karroo-beds. This depression might have formed "als Begleiterscheinung der ältesten Phase der Kapidenfaltung". The phase to which Krenkel refers here (p. 593) is the Upper-Carboniferous. The basin's southern margin was later folded by the Triassic epoch of the Cape Mountains. The latter had probably been much larger during Carboniferous time than the chain's fragment is at present, as part of these ancient belts have foundered in the ocean since then.

Sedimentation in the Karroo Basin consists of the so-called Karroo-system, i.e. a sequence of strata totalling as much as 6,700 m (Krenkel, p. 805). On this rests as much as 1,400 m of volcanic rocks. The sediments range from the carboniferous Dwyka series to the triassic Stormberg-beds. An analogous series is found in the Congo basin (no. 27) but the sediments appear to be far thinner in this last area. This should be attributed to the different ways in which the basins originated. Veatch (who published a memoir on the Congo Basin in 1933) writes that "the present Congo geological basin is not an original basin of deposition, but a basin of subsequent formation". During the deposition of the Middle and Upper Lukuga beds, representing the Permian Ecca and Lower Beaufort-beds of South Africa, the Congo region had a topography of no great relief, sloping in general from NW to SE.

"Following the deposition of the Lower Beaufort (Upper Lukuga) Permian beds, this region was subjected to the disturbances which marked the end of the Permian and continued through the early Triassic, and which on the one hand produced a great east-west mountain range in what is now the ocean south of the Cape, and on the other the uplift, with associated faulting and folding, which resulted in the general absence of the Middle and Upper Beaufort (Lower and Middle Triassic) in Africa north of about the 26th Parallel, except near the present East Coast. During this early Triassic erosion period all the Lower Beaufort, Ecca, and Dwyka beds (that is, the Upper, Middle, and Lower Lukuga of the Congo) were removed by erosion, except where they had been infaulted or infolded.

"Incidents of this period of folding, faulting, and erosion were the production, in the Stanleyville region of the Congo and the Cassange region of Angola, in the eastern and southwestern part of the present Congo Basin respectively, of local fault basins, which may appropriately be called early rift-basins."

The region was subjected to peneplanation which produced successively the mid-Cretaceous peneplain, the Miocene peneplain, and the end-Tertiary peneplain. "The present Congo hydrographic basin is entirely post-Miocene and owes its origin to the uplift and warping of the Miocene peneplain."

The present shape of the Congo basin (like that of the basin of Paris, see Chapter III) differs entirely from its original tectonic design.

The formation of the plate-shaped depression (no. 28) of the Kalahari seems to date from post Upper-Cretaceous time and to contain from tertiary up to recent strata (Krenkel, p. 677). A still younger formation is the basin of theTshad (no. 25). This contains upper-tertiary (miocene? or pliocene, see Krenkel p. 1379), pleistocene and recent deposits. A boring in the late tertiary or so-called Tshad-beds revealed that the latter's thickness amounts to 100 m (Krenkel, p. 1384). Another very young formation is the basin of Ghasal (no. 26); see Krenkel p. 16, 132.

The most northern basins of Asia — the Karasea Basin (no. 30), and that of Tschuktchen (no. 31) — were discussed in Chapter III.

The basins of Tungus and Lena-Wilui are situated in the Angara shield. Cambrian and Silurian were deposited in the basin of Tungus (no. 33), but no devonian or lower-carboniferous strata have been observed here so far, though upper-carboniferous and permian deposits are known to occur in said area. Very faint folding marked the close of the Paleozoic, but the beds on the western side were folded during the caledonian and variscian revolutions. It would consequently seem that a subsiding tendency occurred on two occassions, i.e. one in the Cambrian and a second in the Upper-Carboniferous. Two subsiding movements are known to have affected the basin of Wilui (no. 34). The first began in the Cambrian and continued up to the Permian, with an epoch of folding towards the Upper-Cambrian. The second occurred in the Jurassic (Leuchs 1, p. 68, fig. 6 p. 87 and 167, fig. 54).

One of the most important basins of the Asiatic Caledonides is that of Gobi (no. 40). Its foundation was subjected to intensive folding and forms the continuation of the surrounding chains (Leuchs 2, p. 72). This basin began to subside in the Lower-Cretaceous, after the late Cimmerian epoch. The series of non-marine deposits (totalling approximately 4,000 m, see Leuchs 2, p. 78) continued up to recent times and show a faint laramide unconformity.

A much older basin is that of Minussinsk (no. 35). It was depressed into the arcuate structure of the Caledonian East-Sajan and the Kusnezki-Alatau (Leuchs 1, p. 151, 152). Marine devonian sediments were deposited in the basin after the late Caledonian movements, and were accompanied by continental Old Red, carboniferous and coal-bearing upper-paleozoic strata. These sediments reveal the effects of only faint variscian folding. The carboniferous strata amount to 1,000 m (see Obrutchew p. 442 for the preceding details).

The basin of Urjanchai (no. 36) originated in the Devonian contemporaneously with the Minussinsk basin. Deposition of marine sediments continued up to the Lower-Carboniferous. The later deposits, however, (permian and jurassic), were terrestrial. "Die

Faltungen und Brüche wurden durch periodische Senkung und Druck der Umrahmung geschaffen" (Obrutchew, p. 445).

The basins of Umrutschi (Leuchs 1, p. 123, fig. 42) and Chikuching subsided transversally over the folded chains of the Variscian Tianchan. Lower-permian up to jurassic sediments of neritic up to terrestrial facies were deposited on their basements (these two basins do not appear on Plate 6, as they are too small for insertion).

Type IV must be considered to include the huge basin-shaped depression of Western Siberia (no. 32). This extends parallel to the Ural and in the South joins the northern part of the Kysylkum and of the Lake Aral district (Leuchs 2, p. 19). The foundation of the West-Siberian Basin, which was probably folded during the Variscian epoch, is not exposed. Marine sedimentation is known to have occurred from the Upper-Jurassic up to the Tertiary (Leuchs 2, p. 32).

Too little is known of the history of the basin of Dsungarei (no. 37) to be able to determine with any amount of certainty whether this formation should be regarded as a nuclear basin (as that of Tarim) or a type of basin analogous to that of Gobi. The last alternative seems the most probably one, since remnants of the prolongation of the Sudetic Tarbagatai presumably still protrude here and there. Steep overthrusts bound the basin of Dsungarei towards the south (the Variscian Tianschan) and the north (the Caledonian Altai). This basin presumably originated as such after the process of variscian folding, but this is still uncertain. No mention will be made here of smaller basins such as those of Krasnojarsk, Tschulym and Ubsa Nor, for there are as yet very few data on these formations.

Deposition in the basin of Tarim (no. 39) began in the Permian or Upper-Carboniferous and continues up to this day. The neritic but chiefly terrestrial deposits are presumably extremely thick and show several unconformities and "germanotype" folding.

The subsidence and sedimentation of the massif of Ordos (no. 41) was particularly important in the south-east (basin of Shensi), where a series of terrestrial strata ranging from Permian to Cretaceous accumulated to a thickness of 4,400 m (the subsidence subsequently ceased). Movements occurred after the Permian. The later deposits lie undisturbed and were consequently uninfluenced by the Cimmerian phases affecting the Variscian chains around the massif and resulting in marginal thrusts in the direction of the basin. Analogous marginal faultplanes also frame the basin of Ferghana (no. 38). Its contents of cenomanian up to neogene strata total about 2,000 m (Leuchs 2, p. 136), but it is not yet known when subsidence began, nor is anything known of the sediments' real thickness. This last also applies to the deposits in the Kara Sea (no. 30) and Tschuktchen Massifs (no. 31), for both are largely covered by a shallow sea.

The "red basin of Szechuan" (no. 42) probably ought to be included in this group, for it is framed by folded chains which overthrust during the Cimmerian epoch, while marine cambrian and ordovician strata — though no silurian, devonian or carboniferous sediments — have been shown to be present within its foundation. The permian deposits are not very thick (400 m), but the presence of sediments ranging from lower-triassic to cretaceous and attaining a thickness of 5,000 m has been reported. The basin's contents (including pre-triassic basalt) were either folded towards the close of the Cretaceous (laramide folding) or at an earlier date, i.e. towards the close of the Jurassic. A fainter and older Cimmerian phase was observed in addition to the above movement (Lee p. 174, 235).

Two of the five larger basins in Australia — the Desert (no. 44) and North (no. 43) basins — have been filled with several thousand meters of carboniferous and permian strata (approx. 3,000 m in the Desert basin, cf. Clarke p. 33). These receptacles are quite far removed from the paleozoic belts, but the great Artesian basin (no. 45) occupies the marginal areas of the Variscides as well as part of the marginal zones of the Caledonides, all of which have been peneplained to a considerable degree since their formation. This Artesian basin contains on an average 1,000 m of jurassic and cretaceous sediments, but the latter sometimes total 2,300 m (Andrews, p. 122). The Murray River and Eucla basins (no. 47 and 46), with their miocene and pliocene deposits, are even younger than the preceding ones (David, p. 121).

Teichert recently announced that the North Basin (no. 43) not only contained carboniferous and permian deposits, as was formerly supposed, but that successive mesozoic transgressions had also left deposits behind them in this sector. A continuous sequence of sediments, ranging from upper-cretaceous to upper-tertiary strata, were deposited subsequently. These were only folded very slightly towards the close of the Tertiary, and Teichert consequently regards the North-West and even the Desert basin (no. 44) as part of a geosyncline representing a branch of the pretertiary geosyncline which I reconstructed in the eastern sector of the East-Indian Archipelago. I would like to point out, however, that as far as the Desert basin is concerned (1) the upper-paleozoic sediments observed in this region consist mostly of estuarine and fluvio-glacial deposits, and that only one part of these sediments are composed of marine deposits (Clarke, p. 33); (2) that the sediments were deposited basin-wise, and that the geosynclinal shape of a trough is therefore lacking in this case, and (3) that the sediments have not been affected by folding, and that I consequently find it impossible to follow Teichert when the latter attempts to connect the Desert Basin with the geosynclinal zone in the East-Indies.

Though sedimentation in the North-West Basin continued during the Tertiary, I find it equally impossible to determine its geosynclinal character in this case. For (1) a paleozoic basin of the same type as that of the Desert basin appears to have existed in this area; (2) rare evidence of apparent transgressions have been observed in the Bajocian and Upper-Jurassic, and these do not in themselves constitute an argument in favor of geosynclinal conditions; (3) neocomian strata appear to be absent, and transgressions of Albian, Cenomanian and Santonian seem to be present; a complete series ranging from upper-cretaceous to pliocene deposits devoid of any influence of laramide and miocene compression, was deposited. Hence this area cannot be connected with the above East-Indian belt. Still, Teichert's publication is important in so far that it mentions the occurrence of cretaceous up to tertiary sedimentation in said area, a sedimentation which was subjected to faint folding towards the close of the Tertiary. This "slightly folded" (p. 86) sequence of sediments may possibly explain the conspicuous and relatively intensive negative anomaly of the isostasy observed by Vening Meinesz in this region.

The above enumeration will now be followed by a brief examination of a few deep-sea basins. Born, Leuchs, Lawson, and others are known to have not only regarded the East-Indian basins as young depressions, but to have also held the same view as regards the other basins along the eastern margin of Asia. Leuchs (1, p. 209) and Born (1932, p. 747) both even assume that the arc of the Aleutes and tertiary strata of Alaska and Kamschatka surround a foundered pre-tertiary "massif", part of which is at present buried beneath the sea.

The Mediterranean consists of a series of basins, the largest being the basin of the Baleares (situated between Spain, the Baleares, North-Africa, Sardinia, Corsica and the southern coast of France), the Tyrrhenian basin, the Adriatic and the eastern part of the Mediterranean: the Ionic Sea, which is known to contain several deeper portions (see v. Seidlitz, p. 17, fig. 3). All these areas foundered in comparatively recent times and I cannot therefore endorse the view that part of the Mediterranean represents a remnant of Tethys (cf. v. Seidlitz, p. 57, 58). The recent origin of the basin of the Baleares can be deduced from the abrupt expiration of the folded chains of the Baleares, Pyrenees and Maritime Alps. The missing connections between these alpine chains seem to have foundered. The Baleares basin must have formed after these alpine chains had been folded, and must consequently be at least of post-oligocene age. Fallot (v. Seidlitz p. 35) is of the opinion that the basin was formed in Miocene time. The Tyrrhenean area still lay above sea-level in the Pliocene (Born 1932, p. 711–714) and would thus constitute a basin of very recent date. This was also concluded by Suess (2, p. 349). The trend of the folded chains in the Apennines and Corsica seem to indicate that the Tyrrhenian basin should be classified under Type IV. The comparatively shallow Adriatic may possibly have had the same origin and may perhaps be of the same age (cf. Born p. 713). It cannot be said for the moment whether it is connected with the basin of Po or whether the latter should be regarded as a marginal deep of the Alps. It is equally difficult to estimate the age of the eastern sector of the Mediterranean, which is surrounded by young Alpine chains in the east. Krenkel speaks of the Syrian arc as an "Aussenrand eines breiten Faltenlandes das nun zum grössten Teile versenkt unter dem Mittelmeere ruht" (Afrika, p. 102), and part of the bottom of the Ionic Sea and the Adriatic would represent a spur-shaped branch of the precambrian African continent according to Born (p. 713). Hence the eastern basin (which may presumably be subdivided into a few separate smaller basins) probably belongs to Type IV. Von Seidlitz holds (p. 365, 366) that the Aegean Sea began to form in the Miocene. The subsiding movements, however, occurring along a series of faults, would have been particularly pronounced in the Pliocene and Pleistocene, and the eastern part of the Mediterranean, too, was very probably transformed into a deep-sea area in Miocene time (cf. also Suess 2, p. 353).

The Caribbean Basin and such deep graben as the Bartlett and Anageda troughs probably originated in the Mid-Pliocene contemporaneously with the occurrence of many faults on neighboring islands and folding of several idiogeosynclines in Northern South-America (Orinoco, Maracaibo, Falcon, Madgalena and Bolivar).

With the aid of the character of the sediments and the trend of the folded chains intersecting the present coasts of the respective islands, Rutten showed that many deep-sea regions in the Antilles only formed after Cretaceous time and that some parts even originated quite recently (1934, p. 556–558). Still, the Caribbean can be subdivided into at least five separate basins and troughs (with an additional submarine graben: Bartlett Trough), and marginal deeps extend along the external side of the Antilles. Several preliminary remarks on the origin of these formations and their correlation to geological and gravimetrical data appeared in papers by Hess in 1937 and 1939, but the final publication has not yet gone to press. We will therefore refrain from entering into a detailed examination of this highly interesting region, which is in many respects so similar to the area of the East-Indian Archipelago.

Born (1932, p. 836 and fig. 366 on p. 830) wished to regard the Caribbean as a foundered block known as Antillia, around which the folded chains would have settled later. The facts put forward by Rutten, however, make it doubtful whether this area may be assumed to represent a structurally homogeneous unit.

An analogous and recent area of subsidence is also found inside the so-called arc of the Southern Antilles. The South-Georgia and South-Orkney-Islands, both of which were well-known to Holtedahl, were described thus by this author: „Die grossen Mengen von klastischem zum Teil grobklastischem terrigenen Material, das in diesen ganz schmalen Gebirgsketten vorkommt (das vulkanische Tuffmaterial natürlich nicht mitgerechnet, weil es ja aus der unmittelbaren Nachbarschaft stammen kann) setzen ein Muttergebiet im jetzigen Tiefseegebiet voraus" (1930, p. 56).

I will confine myself to the above examples since I have no intention of attempting to give an exhaustive survey of deep-sea troughs, basins and submarine furrows. Of course, other examples might very easily be added — for instance: the marginal deeps of Eastern Asia, Western India and the Southern Antilles, the relief of the foundered part of Melanesia (indicated schematically on Pl. 6), etc.

All deep-sea basins discussed so far are either situated inside or immediately alongside continental folded chains and island-arcs. It should be remembered, however, that basin shaped structures are also found in the floors of the large oceanic sectors proper. These were dealt with in Chapter VI.

Postscriptum.

The following three books appeared when the manuscript had gone to press:

Furon, R. *La Palaeogéographie. Essai sur l'évolution des continents et des océans* (1941).

Gerth, H. *Geologie Südamerikas* (vol. I, Teil 3, 1941).

Stille, H. *Einführung in die Geologie Amerikas* (1940).

The above works could still be mentioned in the lists of literature at the back of Chapters II and VI. It was impossible, however, to discuss their contents in the text.

Additional material from *The pulse of the earth*,
ISBN 978-94-017-6445-2, is available at http://extras.springer.com

GPSR Compliance

The European Union's (EU) General Product Safety Regulation (GPSR) is a set of rules that requires consumer products to be safe and our obligations to ensure this.

If you have any concerns about our products, you can contact us on ProductSafety@springernature.com

In case Publisher is established outside the EU, the EU authorized representative is:

Springer Nature Customer Service Center GmbH
Europaplatz 3
69115 Heidelberg, Germany

Batch number: 10620854

Printed by Printforce, the Netherlands